Columba Sara Evelyn (Ed.)

Frankley Reservoir

Columba Sara Evelyn (Ed.)

Frankley Reservoir

Reservoir, Drinking water, Birmingham, List of Acts of the Parliament of the United Kingdom, 1880–1899

Fec Publishing

Imprint

All parts of this book are extracted from Wikipedia, the free encyclopedia (www.wikipedia.org).

You can get detailed informations about the authors of this collection of articles at the end of this book. The editors (Ed.) of this book are no authors. They have not modified or extended the original texts.

Pictures published in this book can be under different licences than the GNU Free Documentation License. You can get detailed informations about the authors and licences of pictures at the end of this book.

The content of this book was generated collaboratively by volunteers. Please be advised that nothing found here has necessarily been reviewed by people with the expertise required to provide you with complete, accurate or reliable information. Some information in this book maybe misleading or wrong. The Publisher does not guarantee the validity of the information found here. If you need specific advice (f.e. in fields of medical, legal, financial, or risk management questions) please contact a professional who is licensed or knowledgeable in that area.

Publisher:
Fec Publishing is a trademark of
International Book Market Service Ltd., 17 Rue Meldrum, Beau Bassin, 1713-01 Mauritius
Email: info@bookmarketservice.com
Website: www.bookmarketservice.com

Published in 2011

Printed in: U.S.A., U.K., Germany. This book was not produced in Mauritius.

ISBN: 978-613-7-34103-2

Contents

Articles

References

Frankley Reservoir

Frankley Reservoir	
Location	Birmingham
Coordinates	52°25′14″N 1°59′55″W
Lake type	reservoir
Basin countries	United Kingdom

Frankley Reservoir is a semi-circular reservoir for drinking water in Birmingham, England. Its construction was authorised by the Birmingham Corporation Water Act of 1892. It was built by Birmingham Corporation Water Department to designs by Abram Kellett of Ealing in 1904.[1]

It contains 900000 cubic metres (imp gal) of water received from the Elan Valley Reservoirs,[2] 73 mi (117 km) away, in Wales, which arrives via the Elan aqueduct, by the power of gravity alone, dropping 171 feet (52 m) - an average gradient of 1 in 2,300.

Before 1987 it was leaking 540 litres per second (). In that year Ground-penetrating radar was used successfully to isolate the leaks.[3]

See also

- List of reservoirs and dams in the United Kingdom
- Frankley Water Treatment Works

External links

- History [4]
- grid reference SP002803

References

[1] Penguin Dictionary of Civil Engineering p347 (*Radar*)
[2] Penguin Dictionary of Civil Engineering p347 (*Radar*)
[3] Penguin Dictionary of Civil Engineering p347 (*Radar*)
[4] http://www.miac.org.uk/frankleywater.htm

Reservoir

A **reservoir** (pronounced either REZ-ur-vore or REZ-er-vwar, etymology: from French *réservoir* a "storehouse [1]) or an **artificial lake** is used to store water. Reservoirs may be created in river valleys by the construction of a dam or may be built by excavation in the ground or by conventional construction techniques such as brickwork or cast concrete.

The term reservoir may also be used to describe underground reservoirs such as an oil or water well.

The Jhonghua Dam on the Dahan River in Taoyuan County, Taiwan.

Types

Valley dammed reservoir

A dam constructed in a valley relies on the natural topography to provide most of the basin of the reservoir. Dams are typically located at a narrow part of a valley downstream of a natural basin. The valley sides act as natural walls with the dam located at the narrowest practical point to provide strength and the lowest practical cost of construction. In many reservoir construction projects people have to be moved and re-housed, historical artifacts moved or rare environments relocated. Examples include the temples of Abu Simbel[2] (which were moved before the construction of the Aswan Dam to create Lake Nasser from the Nile in Egypt) and the re-location of the village of Capel Celyn during the construction of Llyn Celyn.[3]

Lake Vyrnwy Reservoir. The dam spans the Vyrnwy Valley and was the first large stone dam built in the United Kingdom.

Construction of a reservoir in a valley will usually necessitate the diversion of the river during part of the build often through a temporary tunnel or by-pass channel.[4]

In hilly regions reservoirs are often constructed by enlarging existing lakes. Sometimes in such reservoirs the new top water level exceeds the watershed height on one or more of the feeder streams such as at Llyn Clywedog in Mid Wales.[5] In such cases additional side dams are required to contain the reservoir.

Stocks Reservoir in Lancashire, England.

Where the topography is poorly suited to a single large reservoir, a number of smaller reservoirs may be constructed in a chain such as in the River Taff valley where the three reservoirs Llwyn-on Reservoir, Cantref Reservoir and Beacons Reservoir form a chain up the valley.[6]

Bank-side reservoir

Where water is taken from a river of variable quality or quantity, bank-side reservoirs may be constructed to store the water pumped or siphoned from the river. Such reservoirs are usually built partly by excavation and partly by the construction of a complete encircling bund or embankment which may exceed 6 km in circumference.[7] Both the floor of the reservoir and the bund must have an impermeable lining or core, often made of puddled clay. The water stored in such reservoirs may have a residence time of several months during which time normal biological processes are able to substantially reduce many contaminants and almost eliminate any turbidity. The use of bank-side reservoirs also allows a water abstraction to be closed down for extended period at times when the river is unacceptably polluted or when flow conditions are very low due to drought. The London water supply system is one example of the use of bank-side storage for all the water taken from the River Thames and River Lee with many large reservoirs such as Queen Mary Reservoir visible along the approach to London Heathrow Airport.[7]

Service reservoir

Service reservoirs[8] store fully treated potable water close to the point of distribution. Many service reservoirs are constructed as water towers, often as elevated structures on concrete pillars where the landscape is relatively flat. Other service reservoirs are entirely underground, especially in more hilly or mountainous country. In the United Kingdom, Thames Water has many underground reservoirs built in the 1800s by the Victorians, most of which are lined with brick. A good example is the Honor Oak Reservoir, constructed between 1901 and 1909. When it was completed it was the largest brick built underground reservoir in the world[9] and is still one of the largest in Europe[10] . The reservoir now forms part of the Southern extension of the Thames Water Ring Main. The top of the reservoir has been grassed over and is now the Aquarias Golf Club[11] .

Service reservoirs perform several functions including ensuring sufficient head of water in the water distribution system and providing hydraulic capacitance in the system to even out peak demand from consumers enabling the treatment plant to run at optimum efficiency. Large service reservoirs can also be managed to so that energy costs in pumping are reduced by concentrating refilling activity at times of day when power costs are low.

History

Five thousand years ago, the craters of extinct volcanoes in Arabia were used as reservoirs by farmers for their irrigation water.[12]

Dry climate and water scarcity in India led to early development of water management techniques, including the building of a reservoir at Girnar in 3000 BC.[13] Artificial lakes dating to the 5th century BC have been found in ancient Greece.[14] An artificial lake in present-day Madhya Pradesh province of India, constructed in the 11th century, covered 650 square metres (7000 sq ft).[13]

In Sri Lanka large reservoirs have been created by ancient Sinhalese kings in order to save the water for irrigation. The famous Sri Lankan king Parākramabāhu I of Sri Lanka stated " do not let a drop of water seep into the ocean without benefiting mankind ". He created the reservoir named Parakrama Samudra(sea of King Parakrama),[15] which has astonished archaeologists.

Uses

Direct water supply

Gibson Reservoir, Montana

Many dammed river reservoirs and most bank-side reservoirs are used to provide the raw water feed to a water treatment plant which delivers drinking water through water mains. The reservoir does not simply hold water until it is needed; it can also be the first part of the water treatment process. The time the water is held for before it is released is known as the *retention time*. This is a design feature that allows particles and silts to settle out, as well as time for natural biological treatment using algae, bacteria and zooplankton that naturally live within the water. However natural limnological processes in temperate climate lakes produces temperature stratification in the water body which tends to partition some elements such as manganese and phosphorus into deep, cold anoxic water during the summer months. In the autumn and winter the lake becomes fully mixed again. During drought conditions, it is sometimes necessary to draw down the cold bottom water and the elevated levels of manganese in particular can cause problems in water treatment plants.[16]

Hydroelectricity

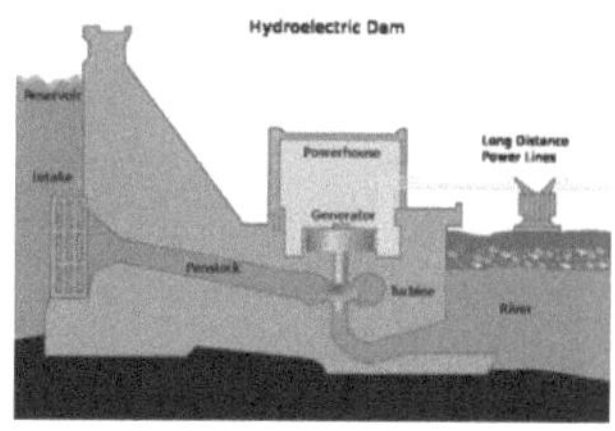

Hydroelectric dam in cross section.

A reservoir generating hydroelectricity includes turbines connected to the retained water body by large-diameter pipes. These generating sets may be at the base of the dam or some distance away. Some reservoirs generating hydroelectricity use pumped re-charge in which a high-level reservoir is filled with water using high-performance electric pumps at times when electricity demand is low and then uses this stored water to generate electricity by releasing the stored water into a low-level reservoir when electricity demand is high. Such systems are called pump-storage schemes.[17]

Controlling watercourses

Reservoirs can be used in a number of ways to control how water flows through downstream waterways.

Downstream water supply – water may be released from an upland reservoir so that it can be abstracted for drinking water lower down the system, sometimes hundred of miles further down downstream

Irrigation – water in an irrigation reservoir may be released into networks of canals for use in farmlands or secondary water systems. Irrigation may also be supported by reservoirs which maintain river flows allowing water to be abstracted for irrigation lower down the river.[18]

Flood control – also known as an "*attenuation*" or "*balancing*" reservoir, flood control reservoirs collect water at times of very high rainfall, then release it slowly over the course of the following weeks or months. Some of these reservoirs are constructed across the river line with the onward flow controlled by an orifice plate. When river flow exceeds the capacity of the orifice plate water builds behind the dam but as soon as the flow rate reduces the water behind the dam slowly releases until the reservoir is empty again. In some cases such reservoirs only function a few times in a decade and the land behind the reservoir may be developed as community or recreational land. A new generation of balancing dams are being developed to combat the climatic consequences of climate change. They are called "Flood Detention Reservoirs". Because these reservoirs will remain dry for long periods, there may be a risk of the clay core drying out reducing its structural stability. Recent developments include the use of composite core fill made from recycled materials as an alternative to clay.

Canals – Where a natural watercourse's water is not available to be diverted into a canal, a reservoir may be built to guarantee the water level in the canal; for example, where a canal climbs to cross a range of hills through locks.[19]

Recreation – water may be released from a reservoir to artificially create or supplement white-water conditions for kayaking and other white-water sports.[20] On salmonid rivers special releases (in Britain called *freshets*) are made to encourage natural migration behaviours in fish and to provide a variety of fishing conditions for anglers.

Recreational-only Kupferbach reservoir near Aachen/Germany.

Flow balancing

Reservoirs can be used to balance the flow in highly managed systems, taking in water during high flows and releasing it again during low flows. In order for this to work without pumping requires careful control of water levels using adjustable sluices. Accurate weather forecasts are essential so that dam operators can plan drawdowns prior to a high rainfall or snowmelt event. Dam operators blamed a faulty weather forecast on the 2010–2011 Queensland floods. Examples of highly managed Reservoirs are Burrendong Dam in Australia and Llyn Tegid in North Wales. Llyn Tegid is a natural lake whose level was raised by a low dam and into which the River Dee flows or discharges depending upon flow conditions at the time as part of the River Dee regulation system. This mode of operation is a form of hydraulic capacitance in the river system.

Recreation

The water bodies provided by many reservoirs often allow some recreational uses such as fishing, boating, and other activities. Special rules may apply for the safety of the public and to protect the quality of the water and the ecology of the surrounding area. Many reservoirs now support and encourage less informal and less structured recreation such as natural history, bird watching, landscape painting, walking and hiking and often provide information boards and interpretation material to encourage responsible use.

Operation

Water falling as rain upstream of the reservoir together with any groundwater emerging as springs is stored in the reservoir. Any excess water can be spilled via a specifically designed spillway. Stored water may be piped by gravity for use as drinking water, to generate hydro-electricity or to maintain river flows to support downstream uses. Occasionally reservoirs can be managed to retain high rain-fall events to prevent or reduce downstream flooding. Some reservoirs support several uses and the operating rules may be complex.

Most modern reservoirs have a specially designed draw-off tower that can discharge water from the reservoir at different levels both to access water as the reservoir draws down but also to allow water of a specific quality to be discharged into the downstream river as compensation water.

The operators of many upland or in-river reservoirs have obligations to release water into the downstream river to maintain river quality, support fisheries, maintain downstream industrial uses. maintain recreational use or for a range of other requirements. Such releases are known as *compensation water.*

Spillway of Llyn Brianne dam in Wales.

Terminology

The terminology for reservoirs varies from country to country. In most of the world reservoir areas are expressed in km^2 whilst in the USA acres are commonly used. For volume either m^3 or km^3 are widely used with acre feet used in the USA.

The capacity, volume or storage of a reservoir is usually divided into distinguishable areas. *Dead* or *inactive* storage refers to water in a reservoir that cannot be drained by gravity through a dam's outlet works, spillway or power plant intake and can only be pumped out. Dead storage allows sediments to settle which improves water quality and also creates hydraulic head along with an area for fish during low levels. *Active* or *live* storage is the portion of the reservoir that can be utilized for flood control, power production, navigation and downstream releases. In addition, a reservoir's *flood control capacity* is the amount of water it can regulate during flooding. The *surcharge capacity* is the capacity of the reservoir above the spillway crest that cannot be regulated.[21]

In the United States the water below the normal maximum level of a reservoir is called the *conservation pool.*[22]

In the UK *top water level* describes the reservoir full state whist *fully drawn down* describes the minimum retained volume.

Modelling reservoir management

There is a wide variety of software for modelling reservoirs, from the specialist Dam Safety Program Management Tools (DSPMT) to the relatively simple WAFLEX, to integrated models like the Water Evaluation And Planning system (WEAP) that place reservoir operations in the context of system-wide demands and supplies.

Safety

In many countries large reservoirs are closely regulated to try to prevent or minimise failures of containment.[23] [24]

Whilst much of the effort is directed at the dam and its associated structures as the weakest part of the overall structure, the aim of such controls is to prevent an uncontrolled release of water from the reservoir. Reservoir failures can generate huge increases in flow down a river valley with the potential to wash away towns and villages and cause considerable loss of life such as the devastation following the failure of containment at Llyn Eigiau which killed 17 people.[25] (see also List of dam failures)

A notable case of reservoirs being used as an instrument of War involved the British Royal Air Force Dambusters raid on Germany in World War II (codenamed "Operation Chastise" [26]), in which three German reservoir dams were selected to be breached in order to impact on German infrastructure and manufacturing and power capabilities deriving from the Ruhr and Eder rivers. The economic and social impact was derived from the enormous volumes of previously stored water that swept down the valleys wreaking destruction. This raid later became the basis for several films.

Environmental impact

Whole life environmental impact

All reservoirs will have a monetary cost/benefit assessment made before construction to see if the project is worth proceeding with.[27] However, such analysis can often omit the environmental impacts of dams and the reservoirs that they contain. Some impacts such as the greenhouse gas production associated with concrete manufacture are relatively easy to estimate. Other impact on the natural environment and social and cultural effects can be more difficult to assess and to weigh in the balance but identification and quantification of these issues are now commonly required in major construction projects in the developed world [28]

Climate change

Depending upon the circumstances, a reservoir built for hydro-electricity generation can either reduce or increase the net production of greenhouse gases. An **increase** can occur if plant material in the flooded areas decays in an anaerobic environment releasing (methane and carbon dioxide). This apparently counterintuitive position arises because much carbon is released as methane which is approximately 8 time more potent as a greenhouse gas than carbon dioxide [29]

A study for the National Institute for Research in the Amazon found that Hydroelectric reservoirs release a large pulse of carbon dioxide from above-water decay of trees left standing in the reservoirs, especially during the first decade after closing.[30] This elevates the global warming impact of the dams to levels much higher than would occur by generating the same power from fossil fuels.[30] According to the World Commission on Dams report (Dams And Development), when the reservoir is relatively large and no prior clearing of forest in the flooded area was undertaken, greenhouse gas emissions from the reservoir could be higher than those of a conventional oil-fired thermal generation plant.[31] For instance, In 1990, the impoundment behind the Balbina Dam in Brazil (inaugurated in 1987) had over 20 times the impact on global warming than would generating the same power from fossil fuels, due to the large area flooded per unit of electricity generated.[30]

A **decrease** can occur if the dam is used in place of traditional power generation, since electricity produced from hydroelectric generation does not give rise to any flue gas emissions from fossil fuel combustion (including sulfur dioxide, nitric oxide and carbon monoxide from coal). The Tucurui dam in Brazil (closed in 1984) had only 0.4 times the impact on global warming than would generating the same power from fossil fuels.[30]

Biology

Dams can produce a block for migrating fish, trapping them in one area, producing food and a habitat for various water-birds. They can also flood various ecosystems on land and may cause extinctions.

Human Impact

Dams can severely reduce the amount of water reaching countries downstream of them, causing water stress between the countries, e.g. the Sudan and Egypt, which damages farming businesses in the downstream countries, and reduces drinking water.

Farms and villages, e.g. Ashopton can be flooded by the creation of reservoirs, ruining many livelihoods. For this very reason, worldwide 80 million people (figure is as of 2009) have had to be forcibly relocated due to dam construction.

Limnology

The limnology of reservoirs has many similarities to that of lakes of equivalent size. There are however significant differences.[32] Many reservoirs experience considerable variations in level producing significant areas that are intermittently underwater or dried out. This greatly limits the productivity or the water margins and limits the number of species able to survive in these conditions.

Upland reservoirs tend to have a much shorter residence time than natural lakes and this can lead to more rapid cycling of nutrients through the water body so that they are more quickly lost to the system. This may be seen as a mismatch between water chemistry and water biology with a tendency for the biological component to be more oligotrophic than the chemistry would suggest.

Conversely, lowland reservoirs drawing water from nutrient rich rivers, may show exaggerated eutrophic characteristics because the residence time in the reservoir is much greater than in the river and the biological systems have a much greater opportunity to utilise the available nutrients.

Deep reservoirs with multiple level draw off towers can discharge deep cold water into the downstream river greatly reducing the size of any hypolimnion. This in turn can reduce the concentrations of phosphorus released during any annual mixing event and may therefore reduce productivity.

The Dams in front of reservoirs act as knickpoints-the energy of the water falling from them reduces and deposition is a result below the Dams.

Seismicity

The filling (impounding) of reservoirs has often been attributed to reservoir-triggered seismicity (RTS) as seismic events have occurred near large dams or within their reservoirs in the past. These events may have been triggered by the filling or operation of the reservoir and are on a small scale when compared to the amount of reservoirs worldwide. Of over 100 recorded events, early examples include the 60 m (197 ft) tall Marathon Dam in Greece (1929), the 221 m (725 ft) tall Hoover Dam in the U.S. (1935). Most events involve large dams and small amounts of seismicity. The only four recorded events above a 6.0-magnitude (M_w) are the 103 m (338 ft) tall Koyna Dam in India which registered a M_w of 6.3 along with the 120 m (394 ft) Kremasta Dam in Greece which registered a 6.3-M_w as well. Following those two, the next largest were the 122 m (400 ft) high Kariba Dam in Zambia at 6.25-M_w and the 105 m (344 ft) Xinfengjiang Dam in China at 6.1-M_w. Disputes occur over when RTS has occurred due to a lack of hydrogeological knowledge at the time of the event. It is accepted though that the infiltration of water into pores and the weight of the reservoir do contribute to RTS patterns. For RTS to occur, there must be a seismic structure near the dam or its reservoir and the seismic structure must be close to failure. Additionally, water must be able to infiltrate the deep rock stratum as the weight of a 100 m (328 ft) deep reservoir will have little impact when compared the deadweight of rock on a crustal stress field which may be located at a depth of 10 km (6 mi) or more.[33]

Micro climate

Reservoirs may change the local micro-climate increasing humidity and reducing extremes of temperature. Such effects are claimed by some South Australian winerys as increasing the quality of the wine production.

List of reservoirs

List of reservoirs by area

The following are the world's ten largest reservoirs by surface area:

1. Lake Volta (8482 km^2 or 3275 sq mi; Ghana) [34]
2. Smallwood Reservoir (6527 km^2 or 2520 sq mi; Canada)[35]
3. Kuybyshev Reservoir (6450 km^2 or 2490 sq mi; Russia)[36]
4. Lake Kariba (5580 km^2 or 2150 sq mi; Zimbabwe, Zambia)[37]
5. Bukhtarma Reservoir (5490 km^2 or 2120 sq mi; Kazakhstan)
6. Bratsk Reservoir (5426 km^2 or 2095 sq mi; Russia)[38]
7. Lake Nasser (5248 km^2 or 2026 sq mi; Egypt, Sudan) [39]
8. Rybinsk Reservoir (4580 km^2 or 1770 sq mi; Russia)
9. Caniapiscau Reservoir (4318 km^2 or 1667 sq mi; Canada)[40]
10. Lake Guri (4250 km^2 or 1640 sq mi; Venezuela)

Lake Volta from space (April 1993).

List of reservoirs by volume

1. Lake Kariba (180 km^3 or 43 cu mi; Zimbabwe, Zambia)
2. Bratsk Reservoir (169 km^3 or 41 cu mi; Russia)
3. Lake Nasser (157 km^3 or 38 cu mi; Egypt, Sudan)
4. Lake Volta (148 km^3 or 36 cu mi; Ghana)
5. Manicouagan Reservoir (142 km^3 or 34 cu mi; Canada)[41]
6. Lake Guri (135 km^3 or 32 cu mi; Venezuela)
7. Williston Lake (74 km^3 or 18 cu mi; Canada)[42]
8. Krasnoyarsk Reservoir (73 km^3 or 18 cu mi; Russia)
9. Zeya Reservoir (68 km^3 or 16 cu mi; Russia)

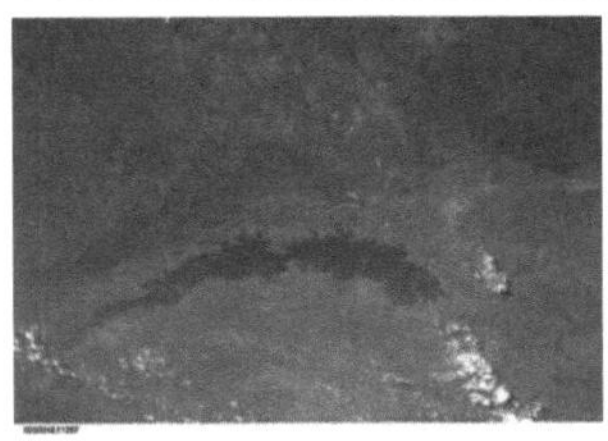

Lake Kariba from space.

See also

- Ab Anbar
- Colourful lakelets (in Poland)
- Drainage basin
- Drought
- Hydroelectricity
- Dam failure
- Mill pond
- Multipurpose reservoir
- Spillway
- Coastal sediment supply

References

[1] Online Etymology Dictionary – Reservoir (http://www.etymonline.com/index.php?term=reservoir)
[2] Nubian Monuments from Abu Simbel to Philae – UNESCO World Heritage Centre (http://whc.unesco.org/en/list/88)
[3] *Capel Celyn, Ten Years of Destruction: 1955–1965*, Thomas E., Cyhoeddiadau Barddas & Gwynedd Council, 2007, ISBN 978 1 900437 92 9
[4] Construction of Hoover Dam: a historic account prepared in cooperation with the Department of the Interior. KC Publications. 1976. ISBN 0-916122-51-4.
[5] Llyn Clywedog – Llanidloes mid-Wales (http://www.llanidloes.com/clywedog/index.html)
[6] Reservoirs of Fforest Fawr Geopark (http://www.geoparcyfforestfawr.org.uk/understanding/archaeology-industrial-heritage/reservoir-of-fforest-fawr-geopark)
[7] Queen Mary and King George V emergency draw down schemes (http://www.icevirtuallibrary.com/content/article/10.1680/dare.2009.19.2.79;jsessionid=3295v7olnvqlv.z-telford-01)
[8] Open University – Service Reservoirs (http://openlearn.open.ac.uk/mod/resource/view.php?id=185924)
[9] "Honor Oak Reservoir" (http://www.lewisham.gov.uk/SiteCollectionDocuments/ForestHillAndHonorOakSecretsLeaflet.pdf). London Borough of Lewisham. . Retrieved 2011-09-01.
[10] "Honor Oak Reservoir" (http://www.projectmanagement.mottmac.com/projects/?mode=type&id=130093). Mott MacDonald. . Retrieved 2011-09-01.
[11] Golf Club website (http://www.aquariusgolfclub.co.uk/pages.php/index.htmlAquarius)
[12] Smith, S. et al. (2006) *Water: the vital resource*, 2nd edition, Milton Keynes, The Open University
[13] edited by John C. Rodda, Lucio Ubertini. (2004). Rodda, John; Ubertini, Lucio. eds. *The Basis of Civilization – Water Science?*. International Association of Hydrological Science. ISBN 1-901502-57-0. OCLC 224463869
[14] Wilson & Wilson (2005). *Encyclopedia of Ancient Greece*. Routledge. ISBN 0415973341. pp. 8
[15] – International Lake Environment Committee – Parakrama Samudra (http://www.ilec.or.jp/database/asi/asi-45.html)
[16] Water problems – Manganese (http://www.freedrinkingwater.com/water_quality/chemical/water-problems-manganese.htm)
[17] How pump storage works (http://www.fhc.co.uk/pumped_storage.htm)
[18] Thinking about an irrigation reservoir? (http://www.ukia.org/eabooklets/EA Reservoir booklet_final.pdf)
[19] Huddersfield narrow canal reservoirs (http://www.huddersfield1.co.uk/huddersfield/narrowcanal/huddscanalres.htm)
[20] Water Release information for The River Tryweryn at the National Whitewater centre (http://www.ukrafting.co.uk/waterinfo.htm)
[21] Vojtěch Broža, Ladislav Votruba, (1989). *Water management in reservoirs* (http://books.google.com/?id=j8dIlPJITH0C&pg=PA187&lpg=PA187&dq=active+storage+reservoir#v=onepage&q=active storage reservoir&f=false). Elsevier Publishing Company. p. 187. ISBN 0-444-98933. .
[22] Lower Colorado River Authority – Water Glossary (http://www.lcra.org/water/conditions/glossary.html)
[23] North Carolina Dam safety law (http://www.dlr.enr.state.nc.us/pages/damsafetylaw1967.html)
[24] Reservoirs Act 1975 The Reservoirs Act 1975 (UK) (http://www.opsi.gov.uk/RevisedStatutes/Acts/ukpga/1975/cukpga_19750023_en_1)
[25] Snowdonia – Llyn Eigau (http://www.snowdoniaguide.com/llyn_eigiau.html)
[26] Commonwealth War Graves Commission – Operation Chastise (http://www.cwgc.org/admin/files/The Dams Raid.pdf)
[27] CIWEM – Reservoirs:Global Issues (http://www.ciwem.org/policy/policies/reservoirs.asp)
[28] Proposed reservoir – Environmental Impact Assessment (EIA) Scoping Report (http://www.whitehorsedc.gov.uk/news_views/topical_issues/xx_detailpage-4402.asp)
[29] Houghton, John (4 May 2005). "Global warming" (http://stacks.iop.org/RoPP/68/1343). *Reports on Progress in Physics* (Institute of Physics) **68** (6): 1362. doi:10.1088/0034-4885/68/6/R02. .
[30] Fearnside, P.M. 1995. hydroelectric dams in the Brazilian Amazon as sources of 'greenhouse' gases. Environmental Conservation 22(1): 7–19.]
[31] Hydroelectric power's dirty secret revealed – earth – 24 February 2005 – New Scientist (http://www.newscientist.com/article.ns?id=dn7046)
[32] Ecology of Reservoirs and Lakes (http://www.forestencyclopedia.net/p/p1483)
[33] "The relationship between large reservoirs and seismicity 08 February 2010" (http://www.waterpowermagazine.com/story.asp?storyCode=2055399). International Water Power & Dam Construction. 20 February 2010. . Retrieved 12 March 2011.
[34] International Lake Environment Committee – Volta Lake (http://www.ilec.or.jp/database/afr/afr-16.html)
[35] The Canadian Encyclopaedia – Smallwood Reservoir (http://www.thecanadianencyclopedia.com/index.cfm?PgNm=TCE&Params=A1ARTA0007465)
[36] International Lake Environment Committee – Reservoir Kuybyshev (http://www.ilec.or.jp/database/eur/deur54.html)
[37] International Lake Environment Committee – Lake Kariba (http://www.ilec.or.jp/database/afr/dafr04.html)
[38] International Lake Environment Committee – Bratskoye Reservoir (http://www.ilec.or.jp/database/asi/dasi61.html)
[39] International Lake Environment Committee – Aswam high dam reservoir (http://www.ilec.or.jp/database/afr/dafr19.html)
[40] International Lake Environment Committee – Caniapiscau Reservoir (http://www.ilec.or.jp/database/nam/dnam35.html)
[41] International Lake Environment Committee – Manicouagan Reservoir (http://www.ilec.or.jp/database/nam/dnam26.html)
[42] International Lake Environment Committee – Williston Lake (http://www.ilec.or.jp/database/nam/dnam29.html)

External links

- Department of Water Resources. "Reservoir Information" (http://cdec.water.ca.gov/misc/resinfo.html). *California Data Exchange Center*. State of California.

ltg:Iudiņture

Drinking water

Drinking water or **potable water** is water pure enough to be consumed or used with low risk of immediate or long term harm. In most developed countries, the water supplied to households, commerce and industry is all of drinking water standard, even though only a very small proportion is actually consumed or used in food preparation. Typical uses include washing or landscape irrigation.

Tap water

Over large parts of the world, humans have inadequate access to potable water and use sources contaminated with disease vectors, pathogens or unacceptable levels of toxins or suspended solids. Such water is not wholesome, and drinking or using such water in food preparation leads to widespread acute and chronic illnesses and is a major cause of death and misery in many countries. Reduction of waterborne diseases is a major public health goal in developing countries.

Water has always been an important and life-sustaining drink to humans and is essential to the survival of all organisms.[1] Excluding fat, water composes approximately 70% of the human body by mass. It is a crucial component of metabolic processes and serves as a solvent for many bodily solutes. The United States Environmental Protection Agency in risk assessment calculations assumes that the average American adult ingests 2.0 litres per day.[2] Drinking water of a variety of qualities is bottled. Bottled water is sold for public consumption throughout the world.

Requirements

Health authorities have historically suggested at least eight glasses, eight fl oz each (240 mL), of water per day (64 fl oz, or 1.89 litres),[3] [2] and the British Dietetic Association recommends 1.8 litres.[1] . This common misconception is not supported by scientific research. Various reviews of all the scientific literature on the topic performed in 2002 and 2008 could not find any solid scientific evidence that recommended drinking eight glasses of water per day.[4] [5] [6] In the US, the reference daily intake (RDI) for water is 3.7 litres per day (L/day) for human males older than 18, and 2.7 L/day for human females older than 18[7] including water contained in food, beverages, and drinking water. The amount of water varies with the individual, as it depends on the condition of the subject, the amount of physical exercise, and on the environmental temperature

Potable water source in the Sahara Desert marked in Italian, German, Polish and English

and humidity.[8] An individual's thirst provides a better guide for how much water they require rather than a specific, fixed quantity.

In terms of mineral nutrients intake, it is unclear what the drinking water contribution is. However, inorganic minerals generally enter surface water and ground water via storm water runoff or through the Earth's crust. Treatment processes also lead to the presence of some minerals. Examples include calcium, zinc, manganese, phosphate, fluoride and sodium compounds.[9] Water generated from the biochemical metabolism of nutrients provides a significant proportion of the daily water requirements for some arthropods and desert animals, but provides only a small fraction of a human's necessary intake. There are a variety of trace elements present in virtually all potable water, some of which play a role in metabolism. For example sodium, potassium and chloride are common chemicals found in small quantities in most waters, and these elements play a role (not necessarily major) in body metabolism. Other elements such as fluoride, while beneficial in low concentrations, can cause dental problems and other issues when present at high levels. Water is essential for the growth and maintenance of our bodies, as it is involved in a number of biological processes.

Profuse sweating can increase the need for electrolyte (salt) replacement. Water intoxication (which results in hyponatremia), the process of consuming too much water too quickly, can be fatal.[10] [11]

Access

Further information: Water resources

Although covering some 70% of the Earth's surface, most water is saline. Freshwater is available in almost all populated areas of the earth, although it may be expensive and the supply may not always be sustainable. Sources where water may be obtained include:

- ground sources such as groundwater, hyporheic zones and aquifers.
- precipitation which includes rain, hail, snow, fog, etc.
- surface water such as rivers, streams, glaciers
- biological sources such as plants.
- the sea through desalination
- Water supply network

Only forty-six percent of people in Africa have safe drinking water.

Spring water, a natural resource from which much bottled water comes, generally contains minerals.[12] Tap water, delivered by domestic water systems in developed nations, refers to water piped to homes through a tap. All of these forms of water are commonly drunk, often purified through filtration.[13]

Drinking water vending machines in Thailand. One litre of purified water is sold (into the customer's own bottle) for 1 baht.

The most efficient way to transport and deliver potable water is through pipes. Plumbing can require significant capital investment. Some systems suffer high operating costs. The cost to replace the deteriorating water and sanitation infrastructure of industrialized countries may be as high as $200 billion a year. Leakage of untreated and treated water from pipes reduces access to water. Leakage rates of 50% are not uncommon in urban systems.[14]

Because of the high initial investments, many less wealthy nations cannot afford to develop or sustain appropriate infrastructure, and as a consequence people in these areas may spend a correspondingly higher fraction of their income on water.[15] 2003 statistics from El Salvador, for example, indicate that the poorest 20% of households spend more than 10% of their total income on water. In the United Kingdom authorities define spending of more than 3% of one's income on water as a hardship.[16]

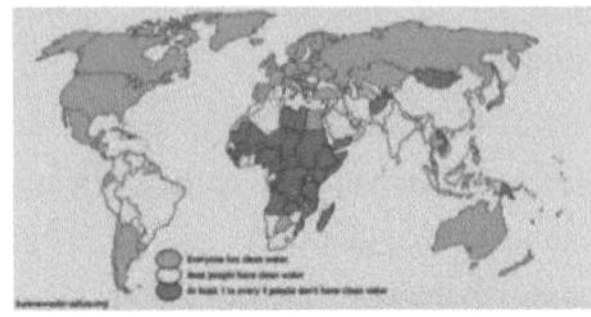

Water quality - percent of population using improved water sources by country

The Millennium Development Goal of halving the proportion of people without access to safe drinking water between 1990 and 2015 will probably be reached. Some countries though still face enormous challenges.[17]

Rural communities are the furthest from meeting the 2015 MDGs drinking water target. Globally only 27% of the rural population has water piped directly to their home and 24% rely on unimproved sources. Of the 884 million people without access to an improved water source, 746 million people (84%) live in rural areas. Sub-Saharan Africa has made the least progress in improved water sources since 1990, improving only 9% to 2006. In contrast, the Eastern Asian region saw a dramatic drop from 45% to 9% reliance on unimproved water in the same time period.[18]

Percentage of population with access to safe drinking water (2000)

Country	%	Country	%	Country	%	Country	%	Country	%
Albania	97	Algeria	89	Azerbaijan	78	Brazil	87	Chile	93
China	75	Cuba	91	Egypt	97	India	84	Indonesia	78
Iran	92	Iraq	85	Kenya	57	North Korea	100	South Korea	92
Mexico	88	Moldova	92	Morocco	80	Mozambique	57	Pakistan	90
Peru	80	Philippines	86	Singapore	100	South Africa	86	Sudan	67
Syria	80	Turkey	82	Uganda	52	Venezuela	83	Zimbabwe	83
Note: All industrialized countries (as listed by UNICEF at 2000) with data available are at 100%.									

In the U.S, the typical nonconserving single family home uses 262 L of water per capita per day. In some parts of the country there are water supplies that are dangerously low due to drought, particularly in the West and the South East region of the U.S.[19]

Improving availability

One of the Millennium Development Goals (MDGs) set by the UN includes environmental sustainability. In 2004, only forty-two percent of people in rural areas had access to clean water.[20]

Solar water disinfection application in Indonesia

Solar water disinfection is a low-cost method of purifying water that can often be implemented with locally available materials.[21] [22] [23] [24] Unlike methods that rely on firewood, it has low impact on the environment.

One organisation working to improve the availability of safe drinking water in some the world's poorest countries is WaterAid International. Operating in 26 countries [25] , WaterAid is working to make lasting improvements to peoples' quality of life by providing long-term sustainable access to clean water in countries such as Nepal, Tanzania, Ghana and India. It also works to educate people about sanitation and hygiene.[26]

The Global Framework for Action (GF4A) is an organization that brings together stakeholders, national governments, donors and NGOs (such as Water aid) to define manageable targets and deadlines. 23 Countries are off-track to meet the MDG goals for improved water availability.[27]

Well contamination

Some efforts at increasing the availability of safe drinking water have been disastrous. When the 1980s was declared the "International Decade of Water" by the United Nations, the assumption was made that groundwater is inherently safer than water from rivers, ponds, and canals. While instances of cholera, typhoid and diarrhea were reduced, other problems emerged. In India for example, 60 million people are estimated to have been poisoned by well water contaminated by excessive fluoride, which is dissolved from the granite rocks. The effects are particularly evident in the bone deformations of children. Similar or larger problems are anticipated in other countries including China, Uzbekistan, and Ethiopia. Although helpful for dental health in low dosage, fluoride in large amounts interferes with bone formation.[28]

In a related problem, it is estimated that half of the Bangladesh's 12 million tube wells contain unacceptable levels of arsenic due to the wells not being dug deep enough (past 100 M). The Bangladeshi government had spent less than $7 million of the 34 million allocated for solving the problem by the World Bank in 1998.[28] [29] Natural arsenic poisoning is a global threat, 140 million people affected in 70 countries on all continents.[30] These examples illustrate the need to examine each location on a case by case basis and not assume what works in one area will work in another.

Diarrhea as a major health effect among children

Over 90% of deaths from diarrheal diseases in the developing world today occur in children under 5 years old. Malnutrition, especially protein-energy malnutrition, can decrease the children's resistance to infections, including water-related diarrheal diseases. From 2000-2003, 769,000 children under five years old in sub-Saharan Africa died each year from diarrheal diseases. As a result of only thirty-six percent of the population in the sub-Saharan region having access to proper means of sanitation, more than 2000 children's lives are lost every day. In South Asia, 683,000 children under five years old died each year from diarrheal disease from 2000-2003. During the same time period, in developed countries, 700 children under five years old died from diarrheal disease. Improved water supply reduces diarrhea morbidity by twenty-five percent and improvements in drinking water through proper storage in the home and chlorination reduces diarrhea episodes by thirty-nine percent.[31]

Water quality and contaminants

Further information: Water quality, Water pollution, and Appropriate technology#Drinking water

Parameters for drinking water quality typically fall under two categories: chemical/physical and microbiological. Chemical/physical parameters include heavy metals, trace organic compounds, total suspended solids (TSS), and turbidity. Microbiological parameters include Coliform bacteria, *E. coli*, and specific pathogenic species of bacteria (such as cholera-causing *Vibrio cholerae*), viruses, and protozoan parasites.

EPA drinking water security poster

Chemical parameters tend to pose more of a chronic health risk through buildup of heavy metals although some components like nitrates/nitrites and arsenic can have a more immediate impact. Physical parameters affect the aesthetics and taste of the drinking water and may complicate the removal of microbial pathogens.

Originally, fecal contamination was determined with the presence of coliform bacteria, a convenient marker for a class of harmful fecal pathogens. The presence of fecal coliforms (like *E. Coli*) serves as an indication of contamination by sewage. Additional contaminants include protozoan oocysts such as *Cryptosporidium sp.*, *Giardia lamblia*, *Legionella*, and viruses (enteric).[32] Microbial pathogenic parameters are typically of greatest concern because of their immediate health risk.

Throughout most of the world, the most common contamination of raw water sources is from human sewage and in particular human faecal pathogens and parasites. In 2006, waterborne diseases were estimated to cause 1.8 million deaths each year while about 1.1 billion people lacked proper drinking water.[33] It is clear that people in the developing world need to have access to good quality water in sufficient quantity, water purification technology and availability and distribution systems for water. In many parts of the world the only sources of water are from small streams often directly contaminated by sewage.

Safety Indicators

Access to safe drinking water is indicated by proper sanitary sources. These improved drinking water sources include household connection, public standpipe, borehole condition, protected dug well, protected spring, and rain water collection. Sources that don't encourage improved drinking water to the same extent as previously mentioned include: unprotected well, unprotected spring, rivers or ponds, vender-provided water, bottled water (consequential of limitations in quantity, not quality of water), and tanker truck water. Access to sanitary water comes hand in hand with access to improved sanitation facilities for excreta. These facilities include connection to public sewer, connection to septic system, pour-flush latrine, and ventilated improved pit latrine. Unimproved sanitation facilities are: public or shared latrine, open pit latrine, or bucket latrine.[34]

Water Treatment

Most water requires some type of treatment before use, even water from deep wells or springs. The extent of treatment depends on the source of the water. Appropriate technology options in water treatment include both community-scale and household-scale point-of-use (POU) designs.[35] A few large urban areas such as Christchurch, New Zealand have access to sufficiently pure water of sufficient volume that no treatment of the raw water is required.[36]

Over the past decade, an increasing number of field-based studies have been undertaken to determine the success of POU measures in reducing waterborne disease. The ability of POU options to reduce disease is a function of both their ability to remove microbial pathogens if properly applied and such social factors as ease of use and cultural appropriateness. Technologies may generate more (or less) health benefit than their lab-based microbial removal performance would suggest.

The current priority of the proponents of POU treatment is to reach large numbers of low-income households on a sustainable basis. Few POU measures have reached significant scale thus far, but efforts to promote and commercially distribute these products to the world's poor have only been under way for a few years.

In emergency situations when conventional treatment systems have been compromised, water borne pathogens may be killed or inactivated by boiling[37] but this requires abundant sources of fuel, and can be very onerous on consumers, especially where it is difficult to store boiled water in sterile conditions and is not a reliable way to kill some encysted parasites such as *Cryptosporidium* or the bacterium *Clostridium*. Other techniques, such as filtration, chemical disinfection, and exposure to ultraviolet radiation (including solar UV) have been demonstrated in an array of randomized control trials to significantly reduce levels of water-borne disease among users in low-income countries,[38] but these suffer from the same problems as boiling methods.

Drinking water regulation

Guidelines for the assessment and improvement of service activities relating to drinking water have been published in the form of International standards for drinking water such as ISO 24510.[39]

European Union

The EU sets legislation on water quality. *Directive 2000/60/EC of the European Parliament and of the Council of 23 October 2000 establishing a framework for Community action in the field of water policy*, known as the water framework directive, is the primary piece of legislation governing water.[40] The Drinking water directive relates specifically to **water intended for human consumption**.

Each member state is responsible for establishing the required policing measures to ensure that the legislation is implemented. For example, in the UK the Water Quality Regulations prescribe maximum values for substances that affect wholesomeness and the Drinking Water Inspectorate polices the water companies.

United States of America

In the United States, the Environmental Protection Agency (EPA) sets standards for tap and public water systems under the Safe Drinking Water Act (SDWA).[41] The Food and Drug Administration (FDA) regulates bottled water as a food product under the Federal Food, Drug, and Cosmetic Act (FFDCA).[42] Bottled water is not necessarily more pure, or more tested, than public tap water.[43] However, there is evidence that the United States federal drinking water regulations do not ensure safe water, as some of the regulations have not been updated with more recent science. Dr. Peter W. Preuss, who became the head of the U.S. EPA's division analyzing environmental risks in 2004, has been "particularly concerned", and has faced controversy in studies which suggest that regulations against certain chemicals should be tightened.[44]

Drinking water preferences of other animals

The qualitative and quantitative aspects of drinking water requirements of domesticated animals are studied and described within the context of animal husbandry. However, relatively few studies have been focused on the drinking behavior of wild animals. A recent study has shown that feral pigeons do not discriminate drinking water according to its content of metabolic wastes, such as uric acid or urea (mimicking faeces- or urine-pollution by birds or mammals respectively).[45]

See also

- Bacteriological water analysis
- Boil-water advisory
- Dual piping
- Food safety
- Groundwater
- Plumbing
- Spragg Bags
- Water fluoridation
- Water purification
- Water security

References

[1] Greenhalgh, Alison (March 2001). "Healthy living - Water" (http://www.bbc.co.uk/health/healthy_living/nutrition/drinks_water.shtml). *BBC Health*. . Retrieved 2007-02-19.

[2] U.S. Environmental Protection Agency (EPA). Dallas, TX (2000-05). "Chapter 3: Exposure Scenario Selection" (http://www.epa.gov/earth1r6/6pd/rcra_c/pd-o/chap3.pdf). . Retrieved 2007-02-19. RCRA Delisting Technical Support Document. p. 8.

[3] "The Benefits of Water" (http://web.archive.org/web/20070116223719/http://www.clevelandclinic.org/health/health-info/docs/2700/2731.asp?index=7250). Cleveland Clinic. Archived from the original (http://www.clevelandclinic.org/health/health-info/docs/2700/2731.asp?index=7250) on 2007-01-16. . Retrieved 2007-02-19.

[4] Research debunks health value of guzzling water (http://www.reuters.com/article/healthNews/idUSN0236679720080402). Reuters, April 2008.

[5] H. Valtin, Drink at least eight glasses of water a day." Really? Is there scientific evidence for "8 × 8"? (http://ajpregu.physiology.org/cgi/content/full/283/5/R993) Am J Physiol Regul Integr Comp Physiol 283: R993-R1004, 2002.

[6] Negoianu, Dan; Goldfarb, Stanley (2008). "Just add water" (http://www.asn-online.org/press/pdf/2008-Media/Water Study.pdf) (PDF). *J. Am. Soc. Nephrol* **19** (6): 1041–1043. doi:10.1681/ASN.2008030274. .

[7] US daily reference intake values (http://iom.edu/Reports/2004/Dietary-Reference-Intakes-Water-Potassium-Sodium-Chloride-and-Sulfate.aspx)

[8] Maton, Anthea bj; Jean Hopkins, Charles William McLaughlin, Susan Johnson, Maryanna Quon Warner, David LaHart, Jill D. Wright (1993). *Human Biology and Health*. Englewood Cliffs, New Jersey, USA: Prentice Hall. ISBN 0-13-981176-1.

[9] World Health Organization (http://www.who.int/en/) (WHO). Geneva, Switzerland. Joyce Morrissey Donohue, Charles O. Abernathy, Peter Lassovszky, George Hallberg. "The contribution of drinking-water to total dietary intakes of selected trace mineral nutrients in the United States." (http://www.who.int/entity/water_sanitation_health/dwq/nutintakes.pdf) Draft, August 2004.

[10] Noakes TD, Goodwin N, Rayner BL, *et al.* (1985). "Water intoxication: a possible complication during endurance exercise" (http://journals.lww.com/acsm-msse/Abstract/1985/06000/Water_intoxication__a_possible_complication_during.12.aspx). *Med Sci Sports Exerc* **17** (3): 370–375. PMID 4021781. .

[11] Noakes TD, Goodwin N, Rayner BL, Branken T, Taylor RK (2005). "Water intoxication: a possible complication during endurance exercise, 1985". *Wilderness Environ Med* **16** (4): 221–7. PMID 16366205.

[12] Schardt, David (2000). "Water, Water Everywhere." (http://www.cspinet.org/nah/water) Center for Science in the Public Interest, Washington, D.C. Accessed 2010-10-26.

[13] Hall, Ellen L.; Dietrich, Andrea M. (2000). "A Brief History of Drinking Water." (http://www.wrb.state.ri.us/program_eohistory.html) Washington: American Water Works Association. Product No. OPF-0051634.

[14] United Nations. World Water Assessment Programme (2009). "Water in a Changing World: Facts and Figures." (http://www.unesco.org/water/wwap/wwdr/wwdr3/pdf/WWDR3_Facts_and_Figures.pdf) World Water Development Report 3. p.5.

[15] (http://news.bbc.co.uk/2/hi/science/nature/7867202.stm) BBC News The water vendors of Nigeria Referenced 2008-10-20

[16] (http://hdr.undp.org/en/media/HDRO6-complete.pdf) page 51 Referenced 2008-10-20

[17] (http://mdgs.un.org/unsd/mdg/Resources/Static/Products/Progress2008/MDG_Report_2008_En.pdf) The Millennium Development Goals Report page 44

[18] (http://mdgs.un.org/unsd/mdg/Resources/Static/Products/Progress2009/MDG_Report_2009_En.pdf) The Millennium Development Goals Report 2009. Referenced 2008-10-20

[19] March 2008, Cashing in on Climate Change (http://www.ibisworld.com/pressrelease/pressrelease.aspx?prid=125:), *IBISWorld*

[20] Africa and the Millennium Development Goals [www.un.org/millenniumgoals/docs/MDGafrica07.pdf]

[21] Conroy, RM.; Meegan, ME.; Joyce, T.; McGuigan, K.; Barnes, J. (Oct 1999). "Solar disinfection of water reduces diarrhoeal disease: an update". *Arch Dis Child* **81** (4): 337–8. doi:10.1136/adc.81.4.337. PMC 1718112. PMID 10490440.

[22] Conroy, R.M.; Meegan, M.E.; Joyce, T.M.; McGuigan, K.G.; Barnes, J. (2001). "Use of solar disinfection protects children under 6 years from cholera". *Arch Dis Child* **85** (4): 293–295. doi:10.1136/adc.85.4.293. PMC 1718943. PMID 11567937.

[23] Rose A, Roy S, Abraham V, Holmgren G, George K, Balraj V, Abraham S, Muliyil J et al. (2006). "Solar disinfection of water for diarrhoeal prevention in Southern India". *Arch Dis Child* **91** (2): 139–141. doi:10.1136/adc.2005.077867. PMC 2082686. PMID 16403847.

[24] Hobbins M. (2003). The SODIS Health Impact Study, Ph.D. Thesis, Swiss Tropical Institute Basel

[25] http://www.wateraid.org/international/what_we_do/where_we_work/default.asp

[26] http://www.wateraid.org

[27] World Health Organization. Global Framework for Action on Sanitation and Water Supply (2009-07-21). "Frequently Asked Questions." (http://www.who.int/entity/water_sanitation_health/GF4A_FAQ_21July09.pdf) Working document.

[28] Pearce, Fred (2006). *When the Rivers Run Dry: Journeys Into the Heart of the World's Water Crisis*. Toronto: Key Porter. ISBN 9781552637418.

[29] Bagla, Pallava (2003-06-05). "Arsenic-Laced Well Water Poisoning Bangladeshis" (http://news.nationalgeographic.com/news/2003/06/0605_030605_arsenicwater.html). *National Geographic News* (Washington: National Geographic Society). .

[30] Bagchi, Sanjit (2007-11-20). "Arsenic Threat Reaching Global Dimensions" (http://www.cmaj.ca/cgi/reprint/177/11/1344.pdf). *Canadian Medical Association Journal* **177** (11): 1344–45. doi:10.1503/cmaj.071456. ISSN 1488-2329. PMC 2072985. PMID 18025421. .

[31] WHO/UNICEF, Water for life: making it happen

[32] EPA. Washington, D.C. "Drinking Water Contaminants: Microorganisms." (http://water.epa.gov/drink/contaminants/index.cfm#Microorganisms) 2010-09-21.

[33] U.S. Centers for Disease Control and Prevention. Atlanta, GA. "Safe Water System: A Low-Cost Technology for Safe Drinking Water." (http://www.cdc.gov/safewater/publications_pages/fact_sheets/WW4.pdf) Fact Sheet, World Water Forum 4 Update. March 2006.

[34] Meeting the MDG Drinking Water and Sanitation Target: A Mid-Term Assessment of Progress [www.who.int/water_sanitation_health/monitoring/jmp04.pdf]

[35] Centre for Affordable Water and Sanitation Technology. Calgary, Alberta. "Household Water Treatment Guide," March 2008. (http://webmail.cawst.org/documents/Camille/New Training Materials/HWT Manual_Mar 08.pdf)

[36] Christchurch City Council. Christchurch, NZ. "Our water - Water supply." (http://www.ccc.govt.nz/homeliving/watersupply/ourwater/index.aspx) Accessed 2010-10-26.

[37] World Health Organization, Geneva (2004). "Guidelines for Drinking-water Quality. Volume 1: Recommendations." (http://www.who.int/water_sanitation_health/dwq/GDWQ2004web.pdf) 3rd ed.

[38] Clasen, T.; Schmidt, W.; Rabie, T.; Roberts, I.; Cairncross, S. (2007-03-12). "Interventions to improve water quality for preventing diarrhea: a systematic review and meta-analysis". *British Medical Journal* **334** (7597): 782. doi:10.1136/bmj.39118.489931.BE. PMC 1851994. PMID 17353208.

[39] ISO 24510 Activities relating to drinking water and wastewater services. Guidelines for the assessment and for the improvement of the service to users

[40] Maria, Kaika (April 2003). "The Water Framework Directive: A New Directive for a Changing Social, Political and Economic European Framework" (http://www.informaworld.com/smpp/content~content=a713666358~db=all). *European Planning Studies*, (Taylor and Francis Group) **11** (3): 299–316. doi:10.1080/09654310303640. . Retrieved 2009-02-10.

[41] Pub.L. 93-523; 42 U.S.C. § 300f (http://www.law.cornell.edu/uscode/42/300f.html) *et seq.* December 16, 1974.

[42] June 25, 1938, ch. 675, 52 Stat. 1040; 21 U.S.C. § 301 (http://www.law.cornell.edu/uscode/21/301.html) *et seq.*

[43] EPA. "Ground water and drinking water - Customer Service." (http://safewater.supportportal.com) Accessed 2010-10-26.

[44] Duhigg, Charles (2009-12-16). "That Tap Water Is Legal but May Be Unhealthy" (http://www.nytimes.com/2009/12/17/us/17water.html?em=&pagewanted=all). *New York Times*: p. A1. .

[45] Olah G, Rózsa L (2006). "Nitrogen metabolic wastes do not influence drinking water preference in feral pigeons" (http://www.zoologia.hu/list/olah_rozsa.pdf). *Acta Zoologica Academiae Scientiarum Hungaricae* **52** (4): 401–406. .

External links

- National Pesticide Information Center - Pesticide in Drinking Water Fact Sheet (http://npic.orst.edu/factsheets/drinkingwater.pdf)
- National Pesticide Information Center - Antimicrobials for Food Processing and Drinking Water Systems (http://npic.orst.edu/ptype/amicrob/food.html)
- Drinking water (http://www.dmoz.org/Science/Environment/Water_Resources/Drinking_Water/) at the Open Directory Project
- U.S. Centers for Disease Control and Prevention (CDC) Healthy Water - Drinking Water (http://www.cdc.gov/healthywater/drinking/) One-stop resource for drinking water including information on tap water, water wells, fluoridation, water testing, water-related diseases and contaminants, etc., plus links to EPA, WHO, and other resources
- US Environmental Protection Agency - National drinking water program (http://www.epa.gov/safewater) - General info, regulations & technical publications
- WHO - Water Sanitation and Health: drinking water quality (http://www.who.int/water_sanitation_health/dwq/en/)
- WHO - Water Sanitation and Health: potabilization systems (http://www.who.int/water_sanitation_health/dwq/wsh0207/en/index4.html)
- The International Water Association (http://www.iwahq.org/)
- Waterwedrink - Links to worldwide drinking water quality websites (http://www.waterwedrink.com/index.htm)
- COVSA (http://www.covsa.org/) American veterans raise money for safe water wells in Vietnam
- UNICEF State of the World's Children 2009 (http://www.unicef.org/sowc09/report/report.php) Full Report with Statistics
- In Historic Vote, UN Declares Water a Fundamental Human Right (http://www.democracynow.org/2010/7/29/in_historic_vote_un_declares_access) - video report by *Democracy Now!*

Birmingham

Birmingham

— City and Metropolitan borough —

From top left: Birmingham City Centre from the west; Selfridges in the Bull Ring; Birmingham Town Hall; St Philip's Cathedral; the University of Birmingham; Alpha Tower.

Coat of Arms of the City Council

Nickname(s): "Brum", "Brummagem", "The Second City", "City of a thousand trades", "Workshop of the World"

Motto: Forward

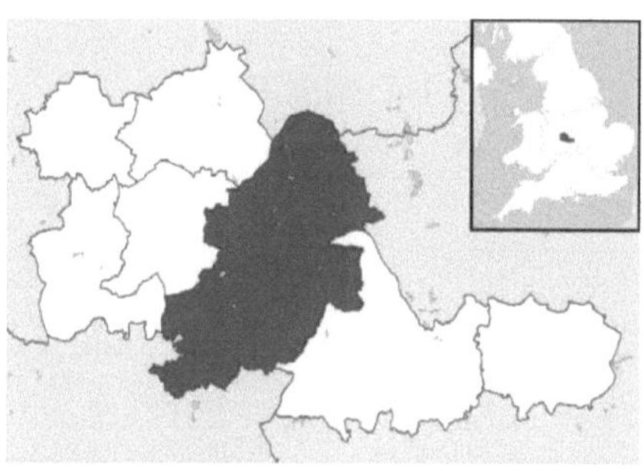

Birmingham shown within England and the West Midlands

Coordinates: 52°28′59″N 1°53′37″W

Sovereign state	United Kingdom

Constituent country	England
Region	West Midlands
Ceremonial county	West Midlands
Admin HQ	The Council House
Founded	7th century
Municipal borough	1838
City	1889
Government	
- Type	Metropolitan borough
- Body	Birmingham City Council
- Lord Mayor	Len Gregory
- Council Leader	Mike Whitby (C)
- Council Control	Conservative / Liberal Democrat Coalition
- MPs	Richard Burden (L) Liam Byrne (L) Jack Dromey (L) Roger Godsiff (L) John Hemming (LD) Khalid Mahmood (L) Shabana Mahmood (L) Steve McCabe (L) Andrew Mitchell (C) Gisela Stuart (L)
Area	
- Total	103.4 sq mi (267.77 km^2)
Elevation	459 ft (140 m)
Population (2008 est.)	
- Total	1,036,900 (Ranked 1st)
- Density	9684/sq mi (3739/km^2)
- Conurbation	2,284,093
- Ethnicity (2007 estimates[1])	66.7% White (62.1% White British) 21.0% South Asian 6.7% Black 3.2% Mixed Race 1.2% Chinese 1.2% Other
Time zone	Greenwich Mean Time (UTC+0)
- Summer (DST)	British Summer Time (UTC+1)
Postcode	B
Area code(s)	0121
ISO 3166 code	GB-BIR
ONS code	00CN
OS grid reference	SP066868
NUTS 3	UKG31

Website	birmingham.gov.uk [2]

Birmingham (🔊 i/ˈbɜrmɪŋəm/ ***bur**-ming-əm*, locally /ˈbɜːmɪŋgəm/ ***bur**-ming-gəm*) is a city and metropolitan borough in the West Midlands county of England. It is the most populous British city outside London, with a population of 1,036,900 (2010 estimate),[3] and lies at the heart of the West Midlands conurbation, the second most populous urban area in the United Kingdom with a population of 2,284,093 (2001 census).[4] Birmingham's metropolitan area, which includes surrounding towns to which it is closely tied through commuting, is also the United Kingdom's second most populous with a population of 3,683,000.[5]

A medium-sized market town during the medieval period, Birmingham grew to international prominence in the 18th century at the heart of the Midlands Enlightenment and subsequent Industrial Revolution, which saw the town at the forefront of worldwide developments in science, technology and economic organisation, producing a series of innovations that laid many of the foundations of modern industrial society.[6] By 1791 it was being hailed as "the first manufacturing town in the world".[7] Birmingham's distinctive economic profile, with thousands of small workshops practising a wide variety of specialised and highly-skilled trades, encouraged exceptional levels of creativity and innovation, and provided a diverse and resilient economic base for an industrial prosperity that was to last into the final quarter of the 20th century.[8] Its resulting high level of social mobility also fostered a culture of broad-based political radicalism, that under leaders from Thomas Attwood to Joseph Chamberlain was to give it a political influence unparalleled in Britain outside London, and a pivotal role in the development of British democracy.[9]

Today Birmingham is a major international commercial centre, ranked as a beta– world city by the Globalization and World Cities Research Network;[10] and an important transport, retail, events and conference hub. With a city GDP of $90bn (2008 estimate, PPP), its urban economy is the second largest in the UK and the 72nd largest in the world.[11] Birmingham's three universities and two university colleges make it the largest centre of higher education in the United Kingdom outside London,[12] and its major cultural institutions, including the City of Birmingham Symphony Orchestra, the Birmingham Royal Ballet and the Barber Institute of Fine Arts, enjoy international reputations.[13] The Big City Plan is a large redevelopment plan currently underway in the city centre with the aim of making Birmingham one of the top 20 most liveable cities in the world within 20 years.[14]

People from Birmingham are known as 'Brummies', a term derived from the city's nickname of 'Brum'. This may originate from the city's dialect name, Brummagem,[15] which may in turn have been derived from one of the city's earlier names, 'Bromwicham'.[16] There is a distinctive Brummie dialect and accent, both of which differ from the adjacent Black Country.

History

Some of the earliest evidence of settlement in Birmingham are artefacts dating back 10,400 years discovered near Curzon Street in the city centre.[17]

Joseph Bailey[18] finds the possible Roman name “Bremium”, for a military station on Icknield Street, not supported by the 14th century original of the *Itinerary* of Richard of Chichester though a later translation has this name. In the early 7th century,[19] Birmingham was an Anglo-Saxon farming hamlet on the banks of the River Rea.[20] The Domesday Book of 1086 introduced the Norman spelling *Bermingeham*, described as a small village, worth 20 shillings.[20] [21] One version of the Saxon name is "Beorma inga ham", meaning farmstead of the sons (or descendants) of Beorma.[20] The other is *Bromwicham*, from the shrub, *broom*, *wych* – village, and *ham* – an abode. *Bromicham* is found in a Halesowen Church Warden's ledger in 1498,[22] and survives in West Bromwich and Castle Bromwich. Leland, 1538, wrote of Bermingham's “pretty street” entering via Deritend.[23] Bailey notes about 150 orthographical changes of the name[24] . *Brummagem* is the common modern nickname.

William Westley's 1731 map of Birmingham. The top of the map is oriented westwards.

In 1166 the holder of the manor of Birmingham, Peter de Birmingham, was granted a royal charter to hold a market in his castle,[17] [25] which in time became known as the Bull Ring, transforming Birmingham from a village to a market town. The de Birmingham family continued to be Lords of Birmingham until the 1530s when Edward de Birmingham was cheated out of its lordship by John Dudley.[26]

As early as the 16th century, Birmingham's access to supplies of iron ore and coal meant that metalworking industries became established.[27] By the time of the English Civil War in the 17th century, Birmingham had become an important manufacturing town with a reputation for producing small arms. Arms manufacture in Birmingham became a staple trade and was concentrated in the area known as the Gun Quarter. During the Industrial Revolution (from the mid-18th century onwards), Birmingham grew rapidly into a major industrial centre and the town prospered. Birmingham's population grew from 15,000 in the late 17th century to 70,000 a century later.[28] During the 18th century, Birmingham was home to the Lunar Society, an important gathering of local thinkers and industrialists.[29]

Birmingham rose to national political prominence in the campaign for political reform in the early nineteenth century, with Thomas Attwood's Birmingham Political Union bringing the country to the brink of civil war and back during the Days of May that preceded the passing of the Great Reform Act in 1832.[30] The Union's meetings on Newhall Hill in 1831 and 1832 were the largest political assemblies Britain had ever seen.[31] Lord Durham, who drafted the act, wrote that "the country owed Reform to Birmingham, and its salvation from revolution".[32]

By the 1820s, an extensive canal system had been constructed, giving greater access to natural resources to fuel to industries. Railways arrived in Birmingham in 1837 with the arrival of the Grand Junction Railway, and a year later, the London and Birmingham Railway. During the Victorian era, the population of Birmingham grew rapidly to well over half a million[33] and Birmingham became the second largest population centre in England. Birmingham was granted city status in 1889 by Queen Victoria.[34] Joseph Chamberlain, who was once mayor of Birmingham and later became an MP, and his son Neville Chamberlain, who was Lord Mayor of Birmingham and later the British Prime Minister, are two of the most well-known political figures who have lived in Birmingham. The city established its own university in 1900.[35]

Birmingham in 1886

Birmingham suffered heavy bomb damage during World War II's "Birmingham Blitz", and the city was extensively redeveloped during the 1950s and 1960s.[36] This included the construction of large tower block estates, such as Castle Vale. The Bull Ring was reconstructed and New Street station was redeveloped.

Birmingham remained by far Britain's most prosperous provincial city as late as the 1970s,[37] with household incomes exceeding even those of London and the South East,[38] but its economic diversity and capacity for regeneration declined in the decades that followed World War II as Central Government sought to restrict the city's growth and disperse industry and population to the stagnating areas of Scotland, Wales and Northern England.[39] These measures hindered "the natural self-regeneration of businesses in Birmingham, leaving it top-heavy with the old and infirm",[40] and the city became increasingly dependent on the motor industry. The recession of the early 1980s saw Birmingham's economy collapse, with unprecedented levels of unemployment and outbreaks of social unrest in inner-city districts.[41] In 1985, the city saw two major outbreaks of civil unrest which had fatal consequences. On 11

May that year, a 14-year-old spectator was crushed to death and many other people were injured in a riot between Birmingham City and Leeds United fans at a Football League game at St Andrew's stadium.[42] Four months later, two people died and dozens more were injured, with extensive damage being caused to vehicles and buildings, during three nights of rioting in the Handsworth district.[43]

In the decades following World War II, the ethnic makeup of Birmingham changed significantly, as it received waves of immigration from the Commonwealth of Nations and beyond.[44] The city's population peaked in 1951 at 1,113,000 residents.[33]

In recent years, Birmingham has been transformed, with the construction of new squares like Centenary Square and Millennium Place. Old streets, buildings and canals have been restored, the pedestrian subways have been removed, and the Bull Ring shopping centre[45] has been completely redeveloped. These were the first steps in the ambitious plans of Birmingham City Council for the redevelopment of Birmingham, which has become known as the Big City Plan.[46]

Government

Birmingham City Council is the largest local authority in the UK and the largest council in Europe[47] with 120 councillors representing 40 wards.[48] Its headquarters are at the Council House in Victoria Square. No single party is in overall control and the council is run by a Conservative/Liberal Democrat coalition led by Mike Whitby.

The Council House, headquarters of Birmingham City Council

The city is also the seat of regional government for the West Midlands region of England as the home of the region's Government Office,[49] the regional development agency Advantage West Midlands,[50] and the West Midlands Regional Assembly.[51]

Birmingham's ten parliamentary constituencies are represented in the House of Commons by one Conservative, one Liberal Democrat and eight Labour MPs.[52] In the European Parliament the city forms part of the West Midlands European Parliament constituency, which elects six Members of the European Parliament.[50]

Birmingham was originally part of Warwickshire, but expanded in the late 19th and early 20th centuries, absorbing parts of Worcestershire to the south and Staffordshire to the north and west. The city absorbed Sutton Coldfield in 1974 and became a metropolitan borough in the new West Midlands county. Up until 1986, the West Midlands County Council was based in Birmingham City Centre.

Law enforcement in Birmingham is carried out by West Midlands Police, whose headquarters are at Lloyd House in Birmingham City Centre. With 87.92 recorded offences per 1000 population in 2009–10, Birmingham's crime rate is above the average for England and Wales, but lower than any of England's other major core cities, and lower than many smaller cities such as Reading, Oxford, Cambridge or Brighton.[53] Fire and rescue services in Birmingham are provided by West Midlands Fire Service and emergency medical care by West Midlands Ambulance Service.

Geography

See also Constituent areas of Birmingham

Birmingham is located in the centre of the West Midlands region of England on the Birmingham Plateau – an area of relatively high ground, ranging around 500 to 1,000 feet (150–300 m) above sea level and crossed by Britain's main north-south watershed between the basins of the Rivers Severn and Trent. To the south west of the city lie the Lickey Hills,[54] Clent Hills and Walton Hill, which reach 1033 feet (315 m) and have extensive views over the city.

The City of Birmingham forms a conurbation with the largely residential borough of Solihull to the south east, and with the city of Wolverhampton and the industrial towns of the Black Country to the north west. Together these make up the West Midlands Urban Area, which covers 59972 ha (600 km^2; 232 sq mi) and has a population of 2,284,093 (2001 Census).[4] Beyond the urban area, Birmingham's metropolitan area – the surrounding area to which it is closely economically tied through commuting – has a population of 3,683,000 (2001 Census) and includes the former Mercian capital of Tamworth and the cathedral city of Lichfield in Staffordshire to the north; the industrial city of Coventry and the Warwickshire towns of Nuneaton, Warwick and Leamington Spa to the east; and the Worcestershire towns of Redditch and Bromsgrove to the south west.[5]

Much of the area now occupied by the city was originally a northern reach of the ancient Forest of Arden, whose former presence can still be felt in the city's dense oak tree-cover and in the large number of districts such as Moseley, Saltley, Yardley, Stirchley and Hockley with names ending in "-ley": the Old English *-lēah* meaning "woodland clearing".[55]

View across the city from the Lickey Hills, with Longbridge in the foreground.

Geology

Geologically, Birmingham is dominated by the *Birmingham Fault* which runs diagonally through the city from the Lickey Hills in the south west, passing through Edgbaston, the Bull Ring to Erdington and Sutton Coldfield in the north east.[56] To the south and east of the fault the ground is largely softer Mercia Mudstone Group (formerly known as Keuper Marl), interspersed with beds of Bunter pebbles and crossed by the valleys of the Rivers Tame, Rea and Cole along with their tributaries.[57] Much of this would have been laid down during the Permian and Triassic periods.[56] To the north and west of the fault, varying from 150 to 600 feet (45–180 m) higher than the surrounding area and underlying much of the city centre, lies a long ridge of harder Keuper Sandstone.[58] [59]

Climate

The climate in Birmingham is classified as a temperate maritime climate, like much of the British Isles, with average maximum temperatures in summer (July) being around 20 °C (68 °F); and in winter (January) around 4.5 °C (40.1 °F). Extreme weather is rare but the city has been known to experience tornados – the most recent being in July 2005 in the south of the city, damaging homes and businesses in the area.[60]

Occasional summer heatwaves, such as the one experienced in July 2006, have become more common in recent years. Winters had become milder since the 1990s with snow becoming much less frequent, although this seems to have been reversed in the last couple of years with the winter of 2009–10 being the coldest for some 30 years. Similar to most other large cities, Birmingham has a considerable 'urban heat island' effect.[61] During the coldest night recorded in Birmingham (14 January 1982), for example, the temperature fell to –20.8 °C (–5.4 °F) at Birmingham Airport on the city's eastern edge, but just –12.9 °C (8.8 °F) at Edgbaston, near the city centre.[62] Relative to other large UK conurbations, Birmingham is a snowy city due to its inland location and comparatively high elevation.[62] Snow showers often pass through the city via the Cheshire gap on North Westerly airstreams, but can also come off the North Sea from North Easterly airstreams.[62]

Environment

There are over 8000 acres (3237 ha) of parkland open spaces in Birmingham.[63] The largest of the parks is Sutton Park covering 2400 acres (971 ha) making it the largest urban nature reserve in Europe.[64] Birmingham Botanical Gardens are a Victorian creation, with a conservatory and bandstand, close to the city centre. The Winterbourne Botanic Garden, maintained by the University of Birmingham, is also located close to the city centre.

Birmingham has many corridors of wildlife that lie in both informal settings such as the Project Kingfisher and Woodgate Valley Country Park and in a selection of parks such as Handsworth Park and Small Heath Park. The City's horticultural training facility at King's Heath Park is paired up with Pershore College. More traditional environmental concerns are constantly raised by volunteer pressure group Birmingham Friends of the Earth. That group advocate sustainable travel such as local rail revival, walking and cycling, reduction in energy demand and waste generally, and the development of environmental technologies in the city.

Demography

Religion	Percentage of population
Buddhist	0.3%
Christian	59%
Hindu	2%
Jewish	0.2%
Muslim	14.3%
Sikh	2.9%
No religion	12.4%
No answer	8.4%

Birmingham is the most populous British city outside London, with 1,028,700 inhabitants according to 2009 estimates.[3] Birmingham's metropolitan area is also the United Kingdom's second most populous with a population of 3,683,000.[5] At the time of the 2001 UK Census, Birmingham's population was 977,087,[65] having fallen since reaching a peak of 1,112,685 in the 1951 Census.[66]

High density canalside apartments in Birmingham City Centre

The population density is 9,451 inhabitants per square mile (3,649/km²) compared to the 976.9 inhabitants per square mile (377.2/km²) for England. Females represented 51.6% of the population whilst men represented 48.4%. More women were 70 or over.[67] 60.4% of the population was aged between 16 and 74, compared to 66.7% in England as a whole.[68]

The ONS estimates that, in 2007, 62.1% of the population was White British, 2.4% White Irish, 2.2% Other White, 21% Asian, 6.7% Black, 1.2% Chinese, 3.2% of mixed race and 1.2% of other ethnic heritage.[69] 57% of primary and 52% of secondary pupils are from non-white British families.[70] 16.5% of the population was born outside the United Kingdom.

60.3% of households were found to be owner occupied and 27.7% were rented from either the city council, housing association or other registered social landlord. The remaining 11.8% of households were rented privately or lived rent free.[68]

The Bimingham Larger Urban Zone, a Eurostat measure of the functional city-region approximated to local government districts, has a population of 2,357,100 in 2004.[71] In addition to Birmingham itself, the LUZ includes the Metropolitan Boroughs of Dudley, Sandwell, Solihull and Walsall, along with the districts of Lichfield, Tamworth, North Warwickshire and Bromsgrove.[72]

Economy

With a city GDP of $90bn (2008 est., PPP), Birmingham has the second-largest urban economy in the United Kingdom and the 72nd-largest in the world.[11] Although the city grew to prominence as a manufacturing and engineering centre, its economy today is dominated by the service sector, which in 2008 accounted for 86% of its employment.[73] Birmingham is the largest centre for employment in public administration, education and health in Great Britain,[74] and after Leeds and Glasgow it is the third-largest centre for employment in banking, finance and insurance outside London.[75] It is ranked as a beta- world city by the Globalization and World Cities Research Network.[10]

Colmore Row in Birmingham's Business District.

Two of Britain's largest banks were founded in Birmingham – Lloyds Bank (now Lloyds Banking Group) in 1765[76] and the Midland Bank (now HSBC Bank) in 1836[77] – as well as Ketley's Building Society, the world's first building society, in 1775.[78] In 2010, Cushman & Wakefield stated that Birmingham was the third best place in the United Kingdom to locate a business, and the 18th best in Europe.[79]

Tourism is also an increasingly important part of the local economy. With major facilities such as the International Convention Centre and National Exhibition Centre the Birmingham area accounts for 42% of the UK conference and exhibition trade.[80] The city's sporting and cultural venues attract large numbers of visitors.

The city's three Universities, (Aston University, University of Birmingham and Birmingham City University) and two University colleges have over 65,000 students and employ around 15,000 staff, making a significant contribution to the city's economy as well as its research and innovation base.

The Jaguar XJ, made by Jaguar Cars at Castle Bromwich Assembly

With an annual turnover of £2.43bn, Birmingham city centre is the UK's third largest retail centre,[81] with the country's busiest shopping centre – the Bullring[82] – and the largest department store outside London – House of Fraser on Corporation Street.[83] The City also has one of only four Selfridges department stores, and the second largest branch of Debenhams in the country.[82] In 2004 the city was ranked as the third best place to shop in the United Kingdom, behind the West End of London and Glasgow, being described as a "world-class shopping centre".[84]

Manufacturing accounts for 10% of employment in Birmingham, a figure below the average for Great Britain as a whole.[73] Despite the decline of manufacturing in the city several significant industrial plants remain, including Jaguar Cars in Castle Bromwich and Cadbury Trebor Bassett in Bournville.

Although the city has seen economic growth greater than the national average in the 21st century[85] the benefits have been uneven, with commuters from the surrounding area obtaining many of the more skilled jobs. The two parliamentary constituencies with the highest unemployment rates in the UK – Ladywood and Sparkbrook and Small Heath – are both in inner-city Birmingham.[86] Growth has also added to stresses on the city's transport. Many major roads and the central New Street railway station operate over capacity at peak times. In 2011 it was announced that Birmingham will become an enterprise zone, which will help small businesses in the region to increase economic growth.[87]

Culture

Music

Birmingham has had a vibrant and varied musical history over the last century. Birmingham bands have made a major contribution to the musical culture of the United Kingdom, with many contemporary bands citing Birmingham bands as a major influence. In the 1960s, the "Brum Beat" era featured blues and early progressive rock bands, such as The Moody Blues and Velvett Fogg. The city is often described as the birthplace of heavy metal music,[88] with pioneering metal bands from the late 1960s and 1970s such as Black Sabbath and Judas Priest, as well as two members of Led Zeppelin, having come from Birmingham. The next decade saw the influential metal band Napalm Death arise from the city.

Black Sabbath, a pioneering band in heavy metal music, was formed in Birmingham.

In the 1970s, members of The Move and The Idle Race formed the Electric Light Orchestra and Wizzard. The 1970s also saw the rise of reggae and ska in the city with such bands as Steel Pulse, UB40, Musical Youth, Beshara and The Beat, expounding racial unity with politically leftist lyrics and multiracial line-ups, mirroring social currents in Birmingham at that time. Seminal 1980s pop band Duran Duran are also from Birmingham.

Ex Cathedra performing at Birmingham Town Hall

Birmingham has also produced a number of popular bands and musicians including Ocean Colour Scene, The Spencer Davis Group, The Streets, and The Twang. Musicians Jeff Lynne, Ozzy Osbourne, Tony Iommi, John Lodge, Roy Wood, Joan Armatrading, Toyah Willcox, Denny Laine, Sukshinder Shinda, Steve Winwood, and Fyfe Dangerfield all grew up in the city.

Jazz has a following in the city, with the Harmonic Festival [89], the Mostly Jazz Festival [90] and the annual International Jazz Festival running alongside the year-round contemporary programme presented by promoters and development agency Birmingham Jazz [91], directed by Tony Dudley-Evans. The musician-led Cobweb Collective [92] also present regular jazz sessions in several venues around the city.

The internationally-renowned City of Birmingham Symphony Orchestra's home venue is Symphony Hall. There is a City Organist; since 1834 only seven men have held this position. The current holder, Thomas Trotter, has been in post since 1983.[93] Weekly recitals have been given since the organ in Birmingham Town Hall was opened.[94]

The Birmingham Triennial Music Festivals took place from 1784 to 1912. Music was specially composed, conducted or performed by Mendelssohn, Gounod, Sullivan, Dvořák, Bantock and Edward Elgar, who wrote four of his most famous choral pieces for Birmingham. Elgar's *The Dream of Gerontius* had its début performance there in 1900. Composers born in the city include Albert William Ketèlbey and Andrew Glover.

Birmingham's other city-centre music venues include The National Indoor Arena, which was opened in 1991, 02 Academy on Bristol Street, which opened in September 2009 replacing the 02 Academy in Dale End, The CBSO Centre, opened in 1997, HMV Institute in Digbeth and the Adrian Boult Hall, which was built along with Paradise Forum and Birmingham Central Library, at Birmingham Conservatoire.

Theatre and performing arts

The Birmingham Hippodrome, home stage of the Birmingham Royal Ballet

Birmingham's leading producing theatre is the Birmingham Repertory Theatre, which was founded by Barry Jackson in 1913 to "serve an art instead of making that art serve a commercial purpose".[95] The Rep pioneered innovations such as the performance of Shakespeare in modern dress,[96] and launched the careers of performers including Laurence Olivier, Ralph Richardson, Peggy Ashcroft, Paul Scofield and Albert Finney.[97] Other theatre companies in Birmingham include the experimental Stan's Cafe, the politically radical Banner Theatre, the Birmingham Stage Company and the Maverick Theatre Company. The Alexandra Theatre and the Birmingham Hippodrome host large-scale touring productions, while professional drama is performed on a wide range of stages across the city, including the Old Rep, the Crescent Theatre, the Custard Factory, the Old Joint Stock Theatre, the Blue Orange Theatre, The Drum in Aston and the *mac* in Cannon Hill Park.

The Birmingham Royal Ballet[98] resides in the city as does the world's oldest vocational dance school, Elmhurst School for Dance.[99]

The Birmingham Opera Company under artistic director Graham Vick has developed an international reputation for its avant-garde productions,[100] which often take place in factories, abandoned buildings and other found spaces around the city.[101] In 2010 it was described by *The Guardian* as "far and away the most powerful example that I've experienced in this country of how and why opera can still matter."[102] More conventional seasons by Welsh National Opera and other visiting opera companies take place regularly at the Birmingham Hippodrome.[103]

Literature

Literary figures associated with Birmingham include Samuel Johnson who stayed in Birmingham for a short period and was born in nearby Lichfield. Arthur Conan Doyle worked in the Aston area of Birmingham whilst poet Louis MacNeice lived in Birmingham for six years. American author Washington Irving produced several of his most famous literary works whilst staying in Birmingham such as *Bracebridge Hall* and *The Humorists, A Medley* which are based on Aston Hall. Influential poets associated with Birmingham include Roi Kwabena, who was the city's sixth poet laureate,[104] and Benjamin Zephaniah, who was born in the city.

Writer W. H. Auden grew up in the Harborne area of the city. Author J. R. R. Tolkien was brought up in Birmingham with many locations in the city such as Moseley bog, Sarehole Mill and Perrott's Folly supposedly being the inspiration for various scenes in The Lord of the Rings. Other famous residents include the award winning political playwright David Edgar.

Birmingham has a vibrant contemporary literary scene, with local authors including David Lodge, Jim Crace, Jonathan Coe, Joel Lane and Judith Cutler.[105] The city's leading literary publisher is the Tindal Street Press, whose authors include prize-winning novelists Catherine O'Flynn, Clare Morrall and Austin Clarke.[106]

Birmingham is the home of the UK's longest-established local science fiction group, launched in 1971 (although there were earlier incarnations in the 1940s and 1960s) and which organises the annual science fiction event Novacon.

Art and design

Rhyl Sands (ca. 1854), Oil on Canvas, by David Cox

The influence of the Royal Birmingham Society of Artists and the Birmingham School of Art made Birmingham an important centre of Victorian art, particularly within the Pre-Raphaelite and Arts and Crafts movements.[107] Major figures included the watercolourist David Cox, whose later works make him an important precursor of impressionism;[108] the Pre-Raphaelite and symbolist Edward Burne-Jones; Walter Langley, the first of the Newlyn School painters;[109] and Joseph Southall, leader of the group of artists and craftsmen known as the Birmingham Group.

The Birmingham Surrealists were among the "harbingers of surrealism" in Britain in the 1930s and the movement's most active members in the 1940s,[110] while more abstract artists associated with the city included Lee Bank-born David Bomberg and CoBrA member William Gear. Birmingham artists were prominent in several post-war developments in art: Peter Phillips was among the central figures in the birth of Pop Art;[111] John Salt was the only major European figure among the pioneers of photo-realism;[112] and the BLK Art Group used painting, collage and multimedia to examine the politics and culture of Black British identity. Contemporary artists from the city include the Turner Prize winner Gillian Wearing and the Turner Prize shortlisted Richard Billingham and John Walker.[113]

Birmingham's role as a manufacturing and printing centre has supported strong local traditions of graphic design and product design. Iconic works by Birmingham designers include the Baskerville font,[114] Ruskin Pottery,[115] the Acme Thunderer whistle,[116] the Art Deco branding of the Odeon Cinemas[117] and the Mini.[118]

Museums and galleries

Barber Institute of Fine Arts

Birmingham has two major public art collections. Birmingham Museum & Art Gallery is best known for its works by the Pre-Raphaelites, a collection "of outstanding importance".[119] It also holds a significant selection of old masters – including major works by Bellini, Rubens, Canaletto and Claude – and particularly strong collections of seventeenth century Italian Baroque painting and English watercolours.[119] Its design holdings include Europe's pre-eminent collections of ceramics and fine metalwork.[119] The Barber Institute of Fine Arts in Edgbaston is one of the finest small art galleries in the world,[120] with a collection of exceptional quality representing Western art from the thirteenth century to the present day.[121]

The council also owns other museums in the city such as Aston Hall, Blakesley Hall, the Museum of the Jewellery Quarter, Soho House, and Sarehole Mill, a popular attraction for fans of J. R. R. Tolkien. The Birmingham Back to Backs are the last surviving court of back-to-back houses in the city.[122] Cadbury World is a museum showing visitors the stages and steps of chocolate production and the history of chocolate and the company. The Ikon Gallery hosts displays of contemporary art, as does Eastside Projects.

Thinktank is Birmingham's main science museum, with an IMAX cinema, a planetarium and a collection that includes the *Smethwick Engine*, the world's oldest working steam engine.[123] Other science-based museums include the National Sea Life Centre in Brindleyplace, the Lapworth Museum of Geology at the University of Birmingham and the Centre of the Earth environmental education centre in Winson Green.

Nightlife and festivals

Nightlife in Birmingham is mainly concentrated along Broad Street and into Brindleyplace. Outside the Broad Street area are many stylish and underground venues. The Medicine Bar in the Custard Factory, hmv Institute, Rainbow Pub and Air are large clubs and bars in Digbeth. In the Chinese Quarter are areas such as the Arcadian and Hurst Street Gay Village, that abound with bars and clubs. Summer Row, The Mailbox, O2 Academy in Bristol Street, St Philips/Colmore Row, St Paul's Square and the Jewellery Quarter all have a vibrant night life. There are a number of late night pubs in the Irish Quarter.[124] Outside the city centre are Moseley and Star City entertainment complex on the former site of Nechells Power Station.[125]

O2 Academy in Bristol Street

Birmingham is home to many national, religious and spiritual festivals including a St. George's Day party. The Birmingham Tattoo is a long-standing military show held annually at the National Indoor Arena. The Caribbean-style Birmingham International Carnival takes place in odd numbered years. Birmingham Pride takes place in the gay village and attracts up to 100,000 visitors each year. From 1997, the city hosted an annual arts festival *ArtsFest*, the largest free arts festival in the UK. In December 2006, the City Council announced that it would no longer hold Artsfest.[126] The city's largest single-day event is its St. Patrick's Day parade (Europe's second largest, after Dublin).[127] Other multicultural events include the Bangla Mela and the Vaisakhi Mela. The Birmingham Heritage Festival is a Mardi Gras style event in August. Caribbean and African culture are celebrated with parades and street performances by buskers.

Other festivals in the city include Moseley Folk Festival (since 2006), which takes place in Moseley private park and mixes new with established folk acts, the Birmingham International Jazz Festival, Birmingham Comedy Festival (since 2001), which has been headlined by such acts as Peter Kay, The Fast Show, Jimmy Carr, Lee Evans and Lenny Henry and Off The Cuff Festival established in 2009. The biennial International Dance Festival Birmingham started in 2008, organised by DanceXchange and involving indoor and outdoor venues across the city. During the Christmas seaon Birmingham is also host to the Frankfurt Christmas Market, held annually since 2001. Modelled on its German counterpart it has become the UK's largest outdoor Christmas market, and is the largest German market outside of Germany and Austria,[128] attracting over 2.8 million visitors in 2009, and 3.1 million the following year proving its popularity.[129]

Architecture

17 & 19 Newhall Street in Birmingham's characteristic Victorian red brick and terracotta

Birmingham is chiefly a product of the 18th, 19th, and 20th centuries; its growth began during the Industrial Revolution. Consequently, relatively few buildings survive from its earlier history, and those that do are protected. There are 1,946 listed buildings in Birmingham and thirteen scheduled ancient monuments.[130] Birmingham City Council also operate a locally listing scheme for buildings that do not fully meet the criteria for statutorily listed status.

Traces of medieval Birmingham can be seen in the oldest churches, notably the original parish church, St Martin in the Bull Ring. A few other buildings from the medieval and Tudor periods survive, among them the *Lad in the Lane*[131] and *The Old Crown*, the 15th century *Saracen's Head* public house and Old Grammar School in Kings Norton[132] and Blakesley Hall.

A number of Georgian buildings survive, including St Philip's Cathedral, Soho House, Perrott's Folly, the Town Hall and much of St Paul's Square. The Victorian era saw extensive building across the city. Major civic buildings such as the Victoria Law Courts (in characteristic red brick and terracotta), the Council House and the Museum & Art Gallery were constructed.[133] St Chad's Cathedral was the first Roman Catholic cathedral to be built in the UK since the Reformation.[134] Across the city, the need to house the industrial workers gave rise to miles of redbrick streets and terraces, many of back-to-back houses, some of which were later to become inner-city slums.[135]

Selfridges, by architects Future Systems.

Postwar redevelopment and anti-Victorianism resulted in the loss of dozens of Victorian buildings like Birmingham New Street Station, and the old Central Library.[136] In inner-city areas too, much Victorian housing was redeveloped. Existing communities were relocated to tower block estates like Castle Vale.[137]

Birmingham City Council now has an extensive tower block demolition and renovation programme. There has been much construction in the city centre in recent years, including the award-winning[138] Future Systems' Selfridges building in the Bullring Shopping Centre, the Brindleyplace regeneration project and the Millennium Point science and technology centre. Funding for many of these projects has come from the European Union; the Town Hall for example received £3 million in funding from the European Regional Development Fund.[139]

Highrise development has slowed since the 1970s and mainly in recent years because of enforcements imposed by the Civil Aviation Authority on the heights of buildings as they could affect aircraft from the Airport (e.g. Beetham Tower).[140]

Transport

Birmingham Airport

Partly because of its inland central location, Birmingham is a major transport hub on the motorway, rail, and canal networks.[141] The city is served by a number of major motorways and probably the best known motorway junction in the UK: Spaghetti Junction.[142]

The National Express UK headquarters are located on Birmingham's Eastside, alongside the newly developed Birmingham Coach Station, which forms the national hub of the company's coach network.

Birmingham Airport, located six miles east of the city centre in the neighbouring borough of Solihull, is the sixth busiest by passenger traffic in the United Kingdom, and the second busiest outside the London area.[143] It is a major base for airlines including Flybe, Ryanair, Bmibaby, Monarch Airlines and Thomson Airways; and is connected by flag carrier airlines to major international hubs including Dubai, New York-Newark, Frankfurt, Paris-Charles de Gaulle and Amsterdam.[144]

Birmingham-built Pendolino at New Street railway station

Local public transport is by bus, local train and tram. Bus routes are mainly operated by National Express West Midlands, which accounts for over 80% of all bus journeys in Birmingham, however, there are around 50 other, smaller registered bus companies.[145] The number 11 outer circle bus routes are the longest urban bus routes in Europe, being 26 miles (42 km) long[146] with 272 bus stops.[147]

The city's main railway station, Birmingham New Street, is the busiest in the United Kingdom outside London, used by over 40.1 million people annually.[148] Birmingham Snow Hill station, another major railway station in the city centre, is also the terminus for the Midland Metro which operates between the station and Wolverhampton, also serving the nearby towns of Bilston, Wednesbury and West Bromwich.[149] Another city centre station, Birmingham Moor Street (its terminal platforms having been restored) became (5th September 2011) the city's third main line station, with express trains to London Marylebone (Chiltern Railways). There are plans to extend the Midland Metro route further into Birmingham city centre.[150] Birmingham has a large rail-based park and ride network that feeds the city centre.

Birmingham is also notable for its extensive canal system, and the city is often noted for having more miles of canal than Venice. The canals fed the industry in the city during the Industrial Revolution. Canalside regeneration schemes such as Brindleyplace have turned the canals into tourist attractions.

Education

University of Birmingham

The city council is England's largest local education authority, directly or indirectly responsible for 25 nursery schools, 328 primary schools, 77 secondary schools[151] and 29 special schools.[152] It also runs the library service, with 4 million visitors annually,[153] and provides around 3,500 adult education courses throughout the year.[154] The main library is Central Library and there are 41 local libraries in Birmingham, plus a regular mobile library service.[155]

Most of Birmingham's state schools are community schools run directly by Birmingham City Council in its role as local education authority (LEA). However, there are a large number of voluntary aided schools within the state system. King Edward's School is perhaps the most prestigious independent school in the city. The seven schools of The King Edward VI Foundation are known nationally for setting very high academic standards and all the schools consistently achieve top positions in national league tables.[156]

Birmingham Central Library

Birmingham is home to three universities: Birmingham City University, Aston University,University of Birmingham;and two university colleges: Newman University College[157] and University College Birmingham.[158] The Birmingham Conservatoire and Birmingham School of Acting, both now part of Birmingham City University, offer higher education in specific arts subjects. The range of Universities and colleges means that there are over 65,000 higher education students in Birmingham, making it the UK's second largest student city to London.

Birmingham Metropolitan College is one of the largest further education colleges in the country,[159] formed through a series of mergers between smaller colleges. Joseph Chamberlain College is the only sixth form college in Birmingham and Solihull to have been awarded both Beacon Status and an overall OFSTED grade 1 (Outstanding).[160]

Since the 1970s, most secondary schools in Birmingham have been 11-16/18 comprehensive schools, while post GCSE students have the choice of continuing their education in either a school's sixth form or at a further education college. Birmingham has always operated a primary school system of 4–7 infant and 7–11 junior schools.

Religion

Statue of Charles Gore, first Bishop of Birmingham, in front of St Philip's Cathedral

Although Christianity remains the largest religion within Birmingham, with 59% of residents stating that they were Christian in the 2001 Census, the city's religious profile is highly diverse: outside London, Birmingham has the United Kingdom's largest Muslim, Sikh and Buddhist communities; its second largest Hindu community; and its seventh largest Jewish community.[161]

St Philip's Cathedral was upgraded from church status when the Anglican Diocese of Birmingham was created in 1905. There are two other cathedrals: St Chad's, seat of the Roman Catholic Archdiocese of Birmingham, and the Greek Orthodox Cathedral of the Dormition of the Mother of God and St Andrew. The Coptic Orthodox Diocese of the Midlands is also based at Birmingham, with a cathedral under construction. The original parish church of Birmingham, St Martin in the Bull Ring, is Grade II* listed. A short distance from Five Ways the Birmingham Oratory was completed in 1910 on the site of Cardinal Newman's original foundation.

The oldest surviving synagogue in Birmingham is the 1825 Greek Revival Severn Street Synagogue, now a Freemason's Lodge hall. It was replaced in 1856 by the Grade II* listed Singers Hill Synagogue. Birmingham Central Mosque, one of the largest in Europe, was constructed in the 1960s.[162] During the late 1990s Ghamkol Shariff Masjid was built in Small Heath. Much more recently Darul Barakaat Mosque was built in the Bordesley Green area by the Ahmadiyya Muslim Community.[163] The Guru Nanak Nishkam Sewak Jatha Sikh Gurdwara was built on Soho Road in Handsworth in the late 1970s and the Buddhist Dhammatalaka Peace Pagoda near Edgbaston Reservoir in the 1990s.

Sport

Aston Villa vs. Birmingham City in the Second City derby at Villa Park.

Birmingham has played an important part in the history of sport. The Football League – the world's first league football competition – was founded by Birmingham resident and Aston Villa director William McGregor, who wrote to fellow club directors in 1888 proposing "that ten or twelve of the most prominent clubs in England combine to arrange home-and-away fixtures each season".[164] The modern game of tennis was developed between 1859 and 1865 by Harry Gem and his friend Augurio Perera at Perera's house in Edgbaston,[165] with the Edgbaston Archery and Lawn Tennis Society remaining the oldest tennis club in the world.[166] The Birmingham and District Cricket League is the oldest cricket league in the world,[167] and Birmingham was the host for the first ever Cricket World Cup, a Women's Cricket World Cup in 1973.[168] Birmingham was the first city to be named National City of Sport by the Sports Council.[169] Birmingham was selected ahead of London and Manchester to bid for the 1992 Summer Olympics,[170] but was unsuccessful in the final selection process, which was won by Barcelona.[171]

Test cricket at Edgbaston Cricket Ground

Today the city is home of two of the country's oldest professional football teams: Aston Villa F.C., who were founded in 1874 and play at Villa Park; and Birmingham City F.C., who were founded in 1875 and play at St Andrew's. Rivalry between the clubs is fierce and the fixture between the two is called the Second City derby.[172] Villa currently play in the Premier League, and have been League champions on seven occasions and European Champions in 1982. Birmingham City currently play in the Championship, the second tier of English football, but are the holders of the Football League Cup. Another Premier League club West Bromwich Albion F.C. play just outside the city boundaries at The Hawthorns.

Six times County Championship winners Warwickshire County Cricket Club play at Edgbaston Cricket Ground, which also hosts test cricket and one day internationals. The venue was the scene of the highest ever score by a batsman in first-class cricket, when Brian Lara scored 501 not out for Warwickshire in 1994.[173] Birmingham has a professional Rugby Union club, Moseley R.F.C., who play at Billesley Common. A second professional club, Birmingham & Solihull R.F.C., play at Damson Park in the neighbouring borough of Solihull.

International athletics at the National Indoor Arena

Two major championship golf courses lie on the city's outskirts. The Belfry near Sutton Coldfield is the headquarters of the Professional Golfers' Association[174] and has hosted the Ryder Cup more times than any other venue.[175] The Forest of Arden Hotel and Country Club near Birmingham Airport is also a regular host of tournaments on the PGA European Tour, including the British Masters and the English Open.[176]

The AEGON Classic is, alongside Wimbledon and Eastbourne, one of only three UK tennis tournaments on the WTA Tour.[177] It is played annually at the Edgbaston Priory Club, which in 2010 announced plans for a multi-million pound redevelopment, including a new showcase centre court and a museum celebrating the game's Birmingham origins.[178]

The Alexander Stadium will be the headquarters of UK Athletics from October 2011.[179] It is one of two British venues to host fixtures in the elite international IAAF Diamond League[180] and is the home of Birchfield Harriers, which has many international athletes amongst its members. The National Indoor Arena hosted the 2007 European Athletics Indoor Championships and 2003 IAAF World Indoor Championships, as well as hosting the annual Aviva Indoor Grand Prix – the only British indoor athletics fixture to qualify as a IAAF Indoor Permit Meeting[181] – and a wide variety of other sporting events. A fifty-metre Olympic sized swimming pool is planned for Ladywood.[182] Professional boxing, hockey, skateboarding, stock-car racing, greyhound racing and speedway also takes place within the city.

Food & drink

The fruit and vegetable section of the Birmingham Wholesale Markets

Birmingham's development as a commercial town was originally based around its market for agricultural produce, established by royal charter in 1166. Despite the industrialisation of subsequent centuries this role has been retained, and the Birmingham Wholesale Markets remain the largest combined wholesale food markets in the country,[183] selling meat, fish, fruit, vegetables and flowers and supplying fresh produce to restaurateurs and independent retailers as far as 100 miles away.[184]

Birmingham is the only English city outside London to have three Michelin starred restaurants: *Simpson's* in Edgbaston, *Turners* in Harborne and *Purnell's* in the city centre.[185]

Birmingham based breweries included Ansells, Davenport's and Mitchells & Butlers.[186] Aston Manor Brewery is currently the only brewery of any significant size. Many fine Victorian pubs and bars can still be found across the city. The oldest inn in Birmingham is the Old Crown in Deritend (circa 1450). The city has a plethora of nightclubs and bars, notably along Broad Street.[187]

The Wing Yip food empire first began in the city and now has its headquarters in Nechells.[188] The Balti, a type of curry, was invented in the city, which has received much acclaim for the 'Balti Belt' or 'Balti Triangle'.[189] Famous food brands that originated in Birmingham include Typhoo tea, Bird's Custard, Cadbury's chocolate and HP Sauce.

Media

The Electric Cinema

Birmingham has several major local newspapers – the daily *Birmingham Mail* the now weekly *Birmingham Post*, and the weekly *Sunday Mercury*, all owned by the Trinity Mirror who also own *What's On* magazine, a fortnightly listings title which has been running for 30 years. *Forward* (formerly *Birmingham Voice*) is a freesheet produced by Birmingham City Council, which is distributed to homes in the city. Birmingham is also the hub for various national ethnic media and the base for two regional Metro editions (east and west Midlands). Birmingham has a long cinematic history.

The Electric Cinema on Station Street is the oldest working cinema in the UK,[190] and Oscar Deutsch opened his first Odeon cinema in Perry Barr during the 1920s. Birmingham-born architect Harry Weedon collaborated with Oscar Deutsch to design over 300 cinemas across the country, most in the distinctive Art Decostyle.[191] An IMAX cinema is located at Millennium Point in the Eastside.[192] Birmingham has also been the location for films including *Felicia's Journey* of 1999, which used locations in Birmingham that were used in *Take Me High* of 1973 to contrast the changes in the city.[193]

The Mailbox, headquarters of BBC Birmingham.

As well as being the location for television dramas, Birmingham is also a national hub for television broadcasting. The BBC has two facilities in the city. The Mailbox, in the city centre, is the location for the national headquarters of BBC English Regions,[194] the headquarters of BBC West Midlands and the BBC Birmingham network production centre, which were previously located at the Pebble Mill Studios in Edgbaston. The BBC Drama Village, based in Selly Oak, is a production facility specialising in television drama.[195]

Central/ATV studios in Birmingham were the location for the recording of many programmes for ITV including *Tiswas* and *Crossroads* until the complex was closed.[196] When Central TV moved to its current Gas Street studios, it was also the main hub for CITV until CITV was moved to Manchester in 2004. All of ITV Central's output from Birmingham now consists of the West and East editions of the regional news programme *Central Tonight*.

The city is served by numerous national and regional radio stations, as well as local radio stations. These include brmb, 102.2 Capital FM Birmingham, Heart West Midlands, Kerrang! Radio, BBC WM, New Style Radio 98.7FM and Smooth Radio's West Midlands News & Admin Team.[197] *The Archers*, the world's longest running radio soap, is recorded in Birmingham for BBC Radio 4.[198]

Science and invention

Matthew Boulton

Birmingham has been the location for some of the most important inventions and scientific breakthroughs. Local inventions and notable firsts include: gas lighting, custard powder, Brylcreem, the magnetron, the first ever use of radiography in an operation,[199] Lewis Paul and John Wyatt's first cotton Roller Spinning machine and the UK's first ever hole-in-the-heart operation, at Birmingham Children's Hospital.[200]

Among the city's notable scientists and inventors are Matthew Boulton, proprietor of the Soho engineering works, Sir Francis Galton, originator of eugenics and important techniques in statistics, Joseph Priestley, chemist and radical and James Watt, engineer and inventor who is associated with the steam engine. Many of these scientists were members of the Lunar Society, which was based in the city.[201]

Twin cities

Birmingham has six twin cities, which Birmingham City Council refers to as "international partner cities".[202] They are:

- Chicago, Illinois, United States
- Frankfurt am Main, Germany[203]
- Johannesburg, South Africa
- Leipzig, Germany[204]
- Lyon, France[205]
- Milan, Italy[206]

There are also Treaties of Friendship between Birmingham and Guangzhou in China,[202] [207] and between Birmingham and Mirpur in Azad Kashmir from where about 90,000 Birmingham citizens originate.[208]

Birmingham, Alabama, USA, is named after the city and shares an industrial kinship.[209]

See also

- Constituent areas of Birmingham
- List of Lord Mayors of Birmingham
- List of people from Birmingham, England
- List of tallest buildings and structures in Birmingham
- Military history of Birmingham

References

Bibliography

- *An History of Birmingham (1783) by William Hutton* [210] at Project Gutenberg
- Gordon E. Cherry (1994). *Birmingham A Study in Geography, History and Planning*. ISBN 0-471-94900-0.
- Canon Doctor Terry Slater (1981). *A History of Warwickshire*. ISBN 0-85033-416-0.
- Johnathan Berg (1994). *Positively Birmingham*. ISBN 0-9523179-0-7.
- A. J. Gerard; Canon Doctor Terry Slater (1996). *Managing a Conurbation: Birmingham and its Region*. ISBN 1-85858-083-8.

Notes

[1] "Resident Population Estimates by Ethnic Group (Percentages)" (http://www.neighbourhood.statistics.gov.uk/dissemination/LeadTableView.do?a=3&b=276800&c=birmingham&d=13&e=13&g=373272&i=1001x1003x1004&m=0&r=1&s=1256075609225&enc=1&dsFamilyId=1812). Neighbourhood Statistics. June 2006. . Retrieved 23 October 2008.

[2] http://www.birmingham.gov.uk/

[3] "Local Authority Profile – Birmingham" (http://www.nomisweb.co.uk/reports/lmp/la/2038431965/report.aspx). *Nomis – Official Labour Market Statistics*. Office for National Statistics. 2011. . Retrieved 10 September 2011.

[4] "Usual resident population: Census 2001, Key Statistics for urban areas" (http://www.statistics.gov.uk/statbase/ssdataset.asp?vlnk=8271&More=Y). Office for National Statistics. . Retrieved 9 June 2007.

[5] "British urban pattern: population data" (http://www.espon.eu/export/sites/default/Documents/Projects/ESPON2006Projects/StudiesScientificSupportProjects/UrbanFunctions/fr-1.4.3_April2007-final.pdf) (PDF). *ESPON project 1.4.3 Study on Urban Functions*. European Union – European Spatial Planning Observation Network. March 2007. pp. 119–120. . Retrieved 19 September 2010.

[6] Uglow, Jenny (2002). *The Lunar Men – the friends who made the future*. London: Faber & Faber. pp. xiii, 500–501. ISBN 0571216102.; Jones, Peter M. (2009). *Industrial Enlightenment: Science, technology and culture in Birmingham and the West Midlands, 1760–1820*. Manchester: Manchester University Press. pp. 14, 19, 71, 82–83, 231–232. ISBN 0719077702.

[7] Hopkins, Eric (1989). *Birmingham: The First Manufacturing Town in the World, 1760–1840*. London: Weidenfeld & Nicolson. p. 26. ISBN 0297794736.

[8] Berg, Maxine (1991). "Commerce and Creativity in Eighteenth-Century Birmingham" (http://books.google.co.uk/books?id=NfkNAAAAQAAJ&pg=PA173). In Berg, Maxine. *Markets and Manufacture in Early Industrial Europe* (http://books.google.com/?id=NfkNAAAAQAAJ). London: Routledge. pp. 174, 184. ISBN 0415037204. . Retrieved 31 January 2011.; Jacobs, Jane (1969). *The economy of cities*. New York: Random House. pp. 86–89. OCLC 5585.

[9] Ward, Roger (2005). *City-state and nation: Birmingham's political history, 1830–1940* (http://books.google.co.uk/books?id=6ocWAQAAIAAJ). Chichester: Phillimore. pp. jacket. ISBN 1860773206. . Retrieved 31 January 2011. "Such a firm basis of local support enabled political leaders, from Thomas Attwood to Neville Chamberlain, to exert leverage in national politics. No provincial city had a greater influence on the shaping of British democracy and on the governance of the British state than Birmingham."; Briggs, Asa (1990) [1965]. *Victorian Cities*. Harmondsworth: Penguin Books. pp. 185; 187-189. ISBN 0140135820.; Jenkins, Roy (2004). *Twelve cities: a personal memoir* (http://books.google.co.uk/books?id=UBmWDk5V9BAC). London: Pan Macmillan. pp. 50-51. ISBN 0330493337. . Retrieved 2011-10-02.

[10] "The World According to GaWC 2010" (http://www.lboro.ac.uk/gawc/world2010t.html). Globalization and World Cities Research Network. . Retrieved 18 September 2011.

[11] Hawksworth, John; Hoehn, Thomas; Tiwari, Anmol (2 November 2009). "Global city GDP rankings 2008–2025" (http://www.ukmediacentre.pwc.com/imagelibrary/detail.aspx?MediaDetailsID=1562) (PDF). *UK Economic Outlook November 2009*. PricewaterhouseCoopers. p. 32. . Retrieved 26 December 2010.

[12] "Table 0 – All students by institution, mode of study, level of study and domicile 2008/09" (http://www.hesa.ac.uk/dox/dataTables/studentsAndQualifiers/download/institution0809.xls?v=1.0). Higher education Statistics Agency. . Retrieved 31 January 2011.

[13] Maddocks, Fiona (6 June 2010). "Andris Nelsons, magician of Birmingham" (http://www.guardian.co.uk/music/2010/jun/06/andris-nelsons-conductor-cbso-city-culture-2013?cat=music&type=article). *The Observer* (Guardian News and Media). . Retrieved 31 January 2011.; Craine, Debra (23 February 2010). "Birmingham Royal Ballet comes of age" (http://entertainment.timesonline.co.uk/tol/arts_and_entertainment/stage/dance/article7036522.ece). *The Times* (Times Newspapers). . Retrieved 31 January 2011.; "The Barber Institute of Fine Arts" (http://www.johansens.com/england/birmingham/the-barber-institute-of-fine-arts). *Johansens*. Condé Nast. . Retrieved 31 January 2011.

[14] "Big City Plan Website" (http://bigcityplan.birmingham.gov.uk/). Birmingham City Council. . Retrieved 27 January 2010.

[15] "Brummagem" (http://www.worldwidewords.org/weirdwords/ww-bru2.htm). Worldwidewords.com. 13 December 2003. . Retrieved 7 June 2008.

[16] William Hutton (1783). *An History of Birmingham*.

[17] "Evidence of Stone Age man found in Digbeth" (http://www.birminghampost.net/news/west-midlands-news/2009/07/24/evidence-of-stone-age-man-found-in-digbeth-65233-24226815/). Birmingham Post. . Retrieved 28 July 2009.

[18] Methodism In Birmingham, Wesleyan-Methodist Magazine, July and August 1863. Published form of a lecture by Rev. Joseph Bailey, a Wesleyan Methodist Minister in Birmingham. Cited sources for origins include Richard of Chichester, Hutton, Swinney, Leland and Nash.

[19] Bassett, Steven (June 2000). "Anglo-Saxon Birmingham" (http://www.ingentaconnect.com/content/maney/mdh/2000/00000025/00000001/art00001;jsessionid=g9v3ls6i9oqu.alexandra). *Midland History* **25**: 1–27. .

[20] "Birmingham or Brummagem?" (http://www.birmingham.gov.uk/cs/Satellite?c=Page&childpagename=SystemAdmin/PageLayout&cid=1223092625903&packedargs=subtype=PressNewsStandardTemplate&website=1&pagename=BCC/Common/Wrapper/Wrapper&rendermode=live). Birmingham City Council. . Retrieved 13 January 2010.

[21] Bailey, *op cit*

[22] Bailey, *op cit*, quoting Nash, *History of Worcestershire*, quoting the church warden's ledger of Hales Owen, "paid … to the organ maker at Bromicham".

[23] Leland, *Itinerary*, 1538, quoted in Bailey, *op cit*

[24] Bailey, *op cit*

[25] "Historic Birmingham" (http://www.birminghamuk.com/historicbirmingham.htm). Birminghamuk.com. . Retrieved 30 May 2009.

[26] "IONA – Birmingham" (http://web.archive.org/web/20071020214707/http://www.project-iona.co.uk/landscape/cityscape_past_and_present/birmingham). Project-iona.co.uk. Archived from the original (http://www.project-iona.co.uk/landscape/cityscape_past_and_present/birmingham) on 20 October 2007. . Retrieved 30 May 2009.

[27] Hugh Miller (1851). *First Impressions of England and Its People*. Gould and Lincoln.

[28] " Birmingham (England, United Kingdom) (http://www.britannica.com/EBchecked/topic/674997/Birmingham)". Encyclopædia Britannica.

[29] "History of Birmingham 2" (http://www.birminghamuk.com/historic2.htm). BirminghamUK. . Retrieved 7 June 2008.

[30] Hilton, Boyd (2006). *A Mad, Bad, and Dangerous People?: England, 1783–1846* (http://books.google.com/?id=D37VByMZSaUC). Oxford University Press. pp. 426–427. ISBN 0-19-822830-9. . Retrieved 7 January 2009.

[31] Flick, Carlos T. (August 1971). "Thomas Attwood, Francis Place, and the Agitation for British Parliamentary Reform". *The Huntington Library Quarterly* (University of California Press) **34** (4): 359. JSTOR 3816950.

[32] Briggs, Asa (1948). "Thomas Attwood and the Economic Background of the Birmingham Political Union". *Cambridge Historical Journal* (Cambridge University Press) **9** (2): 190. JSTOR 3020620.

[33] "Historic Population Of Birmingham" (http://www.birmingham.gov.uk/cs/Satellite?c=Page&childpagename=Lib-Central-Archives-and-Heritage/PageLayout&cid=1223092760414&pagename=BCC/Common/Wrapper/Wrapper). Birmingham City Council. . Retrieved 13 January 2010.

[34] "History of Mayoralty" (http://web.archive.org/web/20080609052542/http://www.birmingham.gov.uk/mayors). Birmingham.gov.uk. Archived from the original (http://www.birmingham.gov.uk/mayors) on 9 June 2008. . Retrieved 7 June 2008.

[35] "Inside the university" (http://web.archive.org/web/20080102065752/http://www.bham.ac.uk/about/). University of Birmingham. Archived from the original (http://www.bham.ac.uk/about/) on 2 January 2008. . Retrieved 7 June 2008.

[36] "1960s Architecture in Birmingham" (http://www.birmingham.gov.uk/cs/Satellite?blobcol=urldata&blobheader=application/pdf&blobheadername1=Content-Disposition&blobheadervalue1=attachment;+filename=6415485.pdf&blobkey=id&blobtable=MungoBlobs&blobwhere=1223418010271&ssbinary=true) (PDF). Birmingham City Council Planning Department. . Retrieved 13 January 2010.

[37] Sutcliffe, Anthony; Smith, Roger (1974). *Birmingham 1939–1970*. History of Birmingham. **3**. London: Oxford University Press. p. 5. ISBN 0-19-215182-7.

[38] Spencer, Ken; Taylor, Andy; Smith, Barbara; Mawson, John; Flynn, Norman; Batley, Richard (1986). *Crisis in the industrial heartland: a study of the West Midlands*. Oxford: Clarendon Press. p. 23. ISBN 0198232691.

[39] Law, Christopher M. (1981). *British Regional Development Since World War I* (http://books.google.com/?id=Dz8OAAAAQAAJ). London: Methuen. p. 47. ISBN 0416323103. . Retrieved 5 February 2011.

[40] Heard, Ian (1989). *Developing Birmingham 1889–1989: 100 years of city planning*. Birmingham: Birmingham City Council Development Department. p. 109. ISBN 0951337114.

[41] Cherry, Gordon E. (1994). *Birmingham: a study in geography, history, and planning*. Belhaven world cities series. Chichester: Wiley. pp. 160–164. ISBN 0-471-94900-0.

[42] "The Dark Shadow of Heysel" (http://bleacherreport.com/articles/321751-the-dark-shadow-of-heysel). Bleacher Report. . Retrieved 2011-09-15.

[43] Cowan, Mark (2011-07-14). "From the Archives: Police parking ticket sowed seeds for riots - CrimeFiles - News" (http://www.birminghammail.net/news/crime-news/2011/07/14/from-the-archives-police-parking-ticket-sowed-seeds-for-riots-97319-29053622/). Birmingham Mail. . Retrieved 2011-09-15.

[44] "Birmingham's Post War Black Immigrants" (http://web.archive.org/web/20080609021432/http://www.birmingham.gov.uk/GenerateContent?CONTENT_ITEM_ID=2392&CONTENT_ITEM_TYPE=0&MENU_ID=10596). Birmingham City Council. Archived from the original (http://www.birmingham.gov.uk/GenerateContent?CONTENT_ITEM_ID=2392&CONTENT_ITEM_TYPE=0&MENU_ID=10596) on 9 June 2008. . Retrieved 22 July 2009.

[45] "Major Developments" (http://www.birmingham.gov.uk/majordevelopments). Birmingham City Council. . Retrieved 7 June 2008.

[46] Bayley, Stephen (30 June 2008). "Article reviewing the Architectural regeneration of Birmingham City Centre" (http://www.guardian.co.uk/artanddesign/2008/jun/29/architecture.regeneration). *Guardian* (UK). . Retrieved 25 January 2010.

[47] "Corporate Marketing and Promotions Team: Sponsorship" (http://www.birmingham.gov.uk/cs/Satellite/sponsorship?packedargs=website=1&rendermode=live). Birmingham City Council. . Retrieved 12 September 2009.

[48] "Wards" (http://www.birmingham.gov.uk/wards). Birmingham City Council. . Retrieved 13 January 2010.

[49] "Contact Us" (http://www.gos.gov.uk/gowm/contactUs/). Government Office for the West Midlands. . Retrieved 3 February 2009.

[50] "Contact us" (http://www.advantagewm.co.uk/about-awm/contact-us/default.aspx). Advantage West Midlands. . Retrieved 3 February 2009.

[51] "Contact Us" (http://www.wmra.gov.uk/contact_us/Contact_Us.aspx). West Midlands Regional Assembly. . Retrieved 3 February 2009.

[52] "Members of Parliament" (http://www.birmingham.gov.uk/mps). Birmingham City Council. . Retrieved 13 January 2010.

[53] "Recorded crime >> Total recorded offences >> 2009–10" (http://www.homeoffice.gov.uk/rds/ia/atlas.html). Home Office. . Retrieved 19 September 2010.

[54] "Lickey Hills Country Park" (http://www.birmingham.gov.uk/lickey). Birmingham City Council. . Retrieved 13 January 2010.

[55] Hooke, Della (2005). "Mercia: Landscape and Environment" (http://books.google.co.uk/books?id=H8Ke9acHFm4C&pg=PA167). In Brown, Michelle P.; Farr, Carol Ann. *Mercia: an Anglo-Saxon kingdom in Europe*. Continuum. p. 167. ISBN 0826477658. . Retrieved 9 May 2009.

[56] Susan Ashby (10 December 2007). "The Geography of Birmingham" (http://web.archive.org/web/20080212130436/http://www.jpservicez-searcharticles.com/article.detail.php/179361/17/Travel/144/Vacations/The_Geography_of_Birmingham). JPServicez Search Articles. Archived from the original (http://www.jpservicez-searcharticles.com/article.detail.php/179361/17/Travel/144/Vacations/The_Geography_of_Birmingham) on 12 February 2008. . Retrieved 24 December 2007.

[57] Victor Skipp (1987). *The History of Greater Birmingham – down to 1830*. Yardley, Birmingham: V. H. T. Skipp. p. 15. ISBN 0-9506998-0-2.

[58] "The Growth of the City, A History of the County of Warwick: Volume 7: The City of Birmingham (1964), pp. 4–25" (http://www.british-history.ac.uk/report.aspx?compid=22959). British History Online. . Retrieved 22 July 2009.

[59] "Solid Geology – 1:250,000 scale (Source: British Geological Survey, NERC)" (http://web.archive.org/web/20061122082116/http://www.defra.gov.uk/erdp/images/wmgifs/wmgeol.GIF) (gif). Department for Environment Food and Rural Affairs. Archived from the original (http://www.defra.gov.uk/erdp/images/wmgifs/wmgeol.GIF) on 22 November 2006. . Retrieved 7 June 2008.

[60] "Birmingham Tornado 2005" (http://web.archive.org/web/20080618215923rn_1/www.birmingham.gov.uk/tornado.bcc). Birmingham City Council. . Retrieved 13 January 2010.

[61] "Inside Out: Living with global warming" (http://www.bbc.co.uk/pressoffice/pressreleases/stories/2007/03_march/27/climate.shtml). BBC. 27 March 2007. . Retrieved 7 June 2008.

[62] Dennis Wheeler; Julian Mayes (1997). *Regional Climates of the British Isles*. Routledge. ISBN 0-415-13930-9.

[63] "Parks and Nature Conservation" (http://www.birmingham.gov.uk/parks). Birmingham City Council. . Retrieved 7 June 2008.

[64] "Introduction to Sutton Park" (http://web.archive.org/web/20080609040835/http://www.birmingham.gov.uk/GenerateContent?CONTENT_ITEM_ID=3039&CONTENT_ITEM_TYPE=0&MENU_ID=1754). Birmingham City Council. Archived

from the original (http://www.birmingham.gov.uk/GenerateContent?CONTENT_ITEM_ID=3039&CONTENT_ITEM_TYPE=0&MENU_ID=1754) on 9 June 2008. . Retrieved 7 June 2008.
[65] "Birmingham" (http://www.statistics.gov.uk/census2001/profiles/00cn.asp). Office for National Statistics. . Retrieved 6 November 2010.
[66] "Birmingham MB/CB: Historical statistics – Population" (http://www.visionofbritain.org.uk/data_cube_page.jsp?data_theme=T_POP&data_cube=N_TOT_POP&u_id=10101001&c_id=10001043&add=N). *A Vision of Britain through Time*. University of Portsmouth. 2009. . Retrieved 6 November 2010.
[67] "2001 Population Census: Gender Profiles" (http://web.archive.org/web/20080609145143/http://www.birmingham.gov.uk/Media?MEDIA_ID=172964) (PDF). Birmingham City Council. Archived from the original (http://www.birmingham.gov.uk/Media?MEDIA_ID=172964) on 9 June 2008. . Retrieved 7 June 2008.
[68] "2001 Census of Population: Key Findings" (http://web.archive.org/web/20080609032140/http://www.birmingham.gov.uk/GenerateContent?CONTENT_ITEM_ID=26205&CONTENT_ITEM_TYPE=0&MENU_ID=11333). Birmingham City Council. Archived from the original (http://www.birmingham.gov.uk/GenerateContent?CONTENT_ITEM_ID=26205&CONTENT_ITEM_TYPE=0&MENU_ID=11333) on 9 June 2008. . Retrieved 7 June 2008.
[69] "Regional Population Estimates by Ethnic Group" (http://neighbourhood.statistics.gov.uk/dissemination/LeadTableView.do?a=3&b=276800&c=birmingham&d=13&e=13&g=373272&i=1001x1003x1004&m=0&r=1&s=1203649391078&enc=1&dsFamilyId=1812). ONS. . Retrieved 22 July 2009.
[70] Graeme Paton (1 October 2007). "One fifth of children from ethnic minorities" (http://www.telegraph.co.uk/news/main.jhtml?xml=/news/2007/09/27/nschools127.xml). *The Daily Telegraph* (UK). . Retrieved 28 March 2008.
[71] "Urban Audit – City Profiles: Birmingham" (http://www.urbanaudit.org/CityProfiles.aspx?CityCode=UK002C&CountryCode=UK). Urban Audit. . Retrieved 5 October 2008.
[72] "Towards a Common Standard" (http://www.london.gov.uk/archive/mayor/economic_unit/docs/wp13_towards_a_common_standard.pdf) (PDF). Greater London Authority. p. 28. . Retrieved 5 October 2008.
[73] "Employee jobs (2008)" (http://www.nomisweb.co.uk/reports/lmp/la/2038431965/report.aspx#tabjobs). *Nomis – official labour market statistics*. Office for National Statistics. . Retrieved 26 December 2010.
[74] "Birmingham – Employee Jobs – Area Comparison – Public administration, education and health employee jobs (Great Britain)" (http://www.nomisweb.co.uk/reports/lmp/la/2038431965/subreports/abi_compared/report.aspx?allInGB=yes&pivot=8&&sort=1&ascending=). *Nomis-Official Labour Market Statistics*. Office for National Statistics. . Retrieved 26 December 2010.
[75] "Birmingham – Employee Jobs – Area Comparison – Banking, finance and insurance employee jobs (Great Britain)" (http://www.nomisweb.co.uk/reports/lmp/la/2038431965/subreports/abi_compared/report.aspx?allInGB=yes&pivot=7&&sort=1&ascending=). *Nomis-Official Labour Market Statistics*. Office for National Statistics. . Retrieved 26 December 2010.
[76] "Lloyds Bank" (http://www.lloydstsb.com/about_ltsb/lloyds_bank.asp). Lloyd's TSB. . Retrieved 7 June 2008.
[77] "The HSBC Group in Europe" (http://web.archive.org/web/20080626020058/http://www.img.ghq.hsbc.com/1/PA_1_1_S5/content/assets/about_hsbc/brief_history_feb2008s3.pdf) (PDF). HSBC. Archived from the original (http://www.img.ghq.hsbc.com/1/PA_1_1_S5/content/assets/about_hsbc/brief_history_feb2008s3.pdf) on 26 June 2008. . Retrieved 7 June 2008.
[78] Rex, Simon (20 April 2010). "The History of Building Societies" (http://www.bsa.org.uk/consumer/factsheets/100009.htm). Building Societies Association. . Retrieved 6 September 2010.
[79] "European Cities Report rates Manchester higher than Birminham" (http://www.birminghampost.net/birmingham-business/birmingham-business-news/other-uk-business/2010/10/05/european-cities-report-rates-manchester-higher-than-birmingham-65233-27400721.pdf). . Retrieved 5 October 2010.
[80] "NEC Group – Conference City" (http://classic-web.archive.org/web/20080401071154/http://www.locatebirmingham.com/conference_city/nec_group). Locate Birmingham. . Retrieved 7 June 2008.
[81] "Retail Footprint 2010 reveals Britain's shopping successes and strugglers" (http://www.caci.co.uk/492.aspx). CACI Ltd. 21 May 2010. . Retrieved 18 July 2010.
[82] "UK's busiest shopping centre" (http://icbirmingham.icnetwork.co.uk/post/news/tm_method=full&objectid=14600178&siteid=50002-name_page.html). icBirmingham. 3 September 2004. . Retrieved 3 February 2008.
[83] "The 10 largest UK Stores, 1998" (http://www.propertymall.com/press/article/3888). *Corporate Intelligence on Retailing – Other – Tesco Overtakes Sainsburys Overall*. propertymall.com. 5 May 1998. . Retrieved 3 February 2008.
[84] Seenan, Gerard (3 March 2004). "Birmingham soars up the shopping list | UK news" (http://www.guardian.co.uk/uk/2004/mar/03/shopping.arts). *The Guardian* (UK). . Retrieved 30 May 2009.
[85] "NUTS3 Gross Value Added (GVA) (1995–2003) Tables" (http://www.statistics.gov.uk/downloads/theme_economy/NUTS3_Tables_1-12.xls) (xls). Office for National Statistics. . Retrieved 7 June 2008.
Data on sheet NUTS33, line 102
[86] "Local area labour markets: statistical indicators – Parliamentary Constituency tables (Unemployment Rate January 2005 to December 2005)" (http://www.statistics.gov.uk/downloads/theme_labour/PCAtables.xls) (xls). Office for National Statistics. . Retrieved 7 June 2008.
[87] "UK Government creates 4 enterprise zones to help small businesses | The Search Office Space Blog | Searchofficespace" (http://news.searchofficespace.com/sos-news/uk-government-create-4-enterprise-zones-to-help-small-businesses.html). News.searchofficespace.com. 2011-07-28. . Retrieved 2011-09-15.

[88] David Konow (2002). *Bang Your Head: The Rise and Fall of Heavy Metal*. Three Rivers Press. ISBN 0-609-80732-3.

[89] http://www.harmonicfestival.co.uk

[90] http://www.mostlyjazz.co.uk

[91] http://www.birminghamjazz.co.uk

[92] http://www.cobwebcollective.com

[93] "Thomas Trotter – Organ" (http://www.patrickgarvey.com/artists/thomas-trotter.html). Patrick Garvey Management. . Retrieved 7 June 2008.

[94] "Birmingham Town Hall: The Organ" (http://web.archive.org/web/20080611231917/http://www.birmingham.gov.uk/GenerateContent?CONTENT_ITEM_ID=37342&CONTENT_ITEM_TYPE=0&MENU_ID=126). Birmingham City Council. Archived from the original (http://www.birmingham.gov.uk/GenerateContent?CONTENT_ITEM_ID=37342&CONTENT_ITEM_TYPE=0&MENU_ID=126) on 11 June 2008. . Retrieved 7 June 2008.

[95] Davies, Andrew (1987). *Other theatres: the development of alternative and experimental theatre in Britain* (http://books.google.co.uk/books?id=YGKvAAAAIAAJ). Totowa, NJ: Barnes & Noble. p. 57. ISBN 0389207063. . Retrieved 2011-09-17.

[96] "Shakespeare & the Birmingham Repertory Theatre 1913-1929" (http://www.str.org.uk/shop/zcart/index.php?main_page=product_info&cPath=1&products_id=16). The Society for Theatre Research. . Retrieved 2011-09-17.

[97] Chambers, Colin (2006). "Birmingham Repertory Theatre" (http://books.google.co.uk/books?id=9PPVN0-SKMwC&pg=PA91). *Continuum Companion to Twentieth Century Theatre*. London: Continuum International Publishing Group. p. 91. . Retrieved 2011-09-17.

[98] "Birmingham Royal Ballet" (http://web.archive.org/web/20080605043954/http://www.birmingham.gov.uk/GenerateContent?CONTENT_ITEM_ID=24788&CONTENT_ITEM_TYPE=0&MENU_ID=12535#ballet). Birmingham City Council. Archived from the original (http://www.birmingham.gov.uk/GenerateContent?CONTENT_ITEM_ID=24788&CONTENT_ITEM_TYPE=0&MENU_ID=12535#ballet) on 5 June 2008. . Retrieved 7 June 2008.

[99] "Elmhurst School for Dance" (http://www.made-schools.org.uk/schools/elmhurst-school-of-dance). MADE. . Retrieved 3 July 2009.

[100] "Birmingham Opera Company" (http://www.artscouncil.org.uk/rfo/birmingham-opera-company/). Arts Council England. 25 September 2009. .

[101] O'Neill, Sinéad (Summer-Autumn 2009). "Getting out of the House: Unorthodox Performance Spaces in Recent British and Irish Productions" (http://muse.jhu.edu/journals/opera_quarterly/v025/25.3-4.o-neill.pdf). *The Opera Quarterly* **25** (3–4): 291. . Retrieved 14 March 2011.

[102] Service, Tom (28 May 2010). "Love opera. Don't fall into the trappings trap" (http://www.guardian.co.uk/music/tomserviceblog/2010/may/27/opera-robert-thicknesse-snobbery). *The Guardian* (Guardian News and Media). . Retrieved 14 March 2011.

[103] "Welsh National Opera" (http://www.artscouncil.org.uk/rfo/welsh-national-opera/). Arts Council England. 25 September 2009. .

[104] "About the Birmingham Poet Laureate" (http://web.archive.org/web/20080609075008/http://www.birmingham.gov.uk/GenerateContent?CONTENT_ITEM_ID=23945&CONTENT_ITEM_TYPE=0&MENU_ID=5247). Birmingham City Council. Archived from the original (http://www.birmingham.gov.uk/GenerateContent?CONTENT_ITEM_ID=23945&CONTENT_ITEM_TYPE=0&MENU_ID=5247) on 9 June 2008. . Retrieved 7 June 2008.

[105] "Literary connections with Birmingham, Warwickshire" (http://www3.shropshire-cc.gov.uk/birmingh.htm). *Literary Heritage West Midlands*. Shropshire Council. October 2009. . Retrieved 24 September 2010.

[106] "Tindal Street Press Celebrates 10 Years" (http://www.booktrade.info/index.php/showarticle/23388). booktrade,info. 24 September 2009. . Retrieved 25 September 2010.

[107] Hartnell, Roy (1996). *Pre-Raphaelite Birmingham*. Studley: Brewin Books. pp. 1–3. ISBN 1858580641.

[108] Pillement, Georges (1978). "The Precursors of Impressionism". In Sérullaz, Maurice. *Phaidon Encyclopedia of Impressionism*. Oxford: Phaidon. p. 39. ISBN 0714818976.

[109] Fox, Caroline; Greenacre, Francis (1985). *Painting in Newlyn 1880–1930*. London: Barbican Art Gallery. p. 8. ISBN 0946372101.

[110] Remy, Michel (2001). *Surrealism in Britain*. London: Lund Humphries. pp. 36, 220 & 284. ISBN 0853318255

[111] Livingstone, Marco (2000). *Pop Art: A Continuing History*. London: Thames & Hudson. p. 172. ISBN 0500282404.

[112] Perks, Jon. "Picture Perfect" (http://iccoventry.icnetwork.co.uk/0825lifestyle/cityliving/profile/tm_headline=picture-perfect&method=full&objectid=14036618&siteid=50002-name_page.html). *Coventry Evening Telegraph*. . Retrieved 7 October 2007.

[113] "The Turner Prize: Year by Year" (http://www.tate.org.uk/britain/turnerprize/history/yearbyyear.shtm). Tate Britain. . Retrieved 19 September 2010.

[114] "John Baskerville of Birmingham" (http://web.archive.org/web/20071012100752/http://www.birmingham.gov.uk/GenerateContent?CONTENT_ITEM_ID=76113&CONTENT_ITEM_TYPE=0&MENU_ID=260). Birmingham City Council. Archived from the original (http://www.birmingham.gov.uk/GenerateContent?CONTENT_ITEM_ID=76113&CONTENT_ITEM_TYPE=0&MENU_ID=260) on 12 October 2007. . Retrieved 29 September 2007.

[115] "Ruskin pottery centenary exhibition" (http://www.geffrye-museum.org.uk/_download/CCECMYVI.pdf) (PDF). The Geffrye Museum, London. . Retrieved 29 September 2007.

[116] "History of the Whistle" (http://www.drcdurham.ca/htm/history_of_the_whistle.htm). District Referee Coordinator – Durham. . Retrieved 29 September 2007.

[117] Jonathan Glancey (18 May 2002). "The mogul's monuments – How Oscar Deutsch's Odeon cinemas taught Britain to love modern architecture" (http://arts.guardian.co.uk/critic/feature/0,1169,717532,00.html). *Guardian* (UK). . Retrieved 29 September 2007.

[118] "Alec Issigonis, Automotive Designer (1906–1988)" (http://www.designmuseum.org/design/alec-issigonis). Design Museum, London. . Retrieved 29 September 2007.
[119] Fisher, Mark (2005). "Birmingham Museum and Art Gallery". *Britain's Best Museums and Galleries: From the Greatest Collections to the Smallest Curiosities*. Harmondsworth: Penguin Books. pp. 208–210. ISBN 0141019603.
[120] "The Barber Institute of Fine Arts" (http://www.johansens.com/england/midlands/the-barber-institute-of-fine-arts). *Johansens*. Condé Nast. . Retrieved 20 September 2010.
[121] Fisher, Mark (2005). "Barber Institute of Fine Arts". *Britain's Best Museums and Galleries: From the Greatest Collections to the Smallest Curiosities*. Harmondsworth: Penguin Books. pp. 205–207. ISBN 0141019603.
[122] "Back to back in Birmingham" (http://www.birmingham.gov.uk/ELibrary?E_LIBRARY_ID=93&a=1157290087529) (PDF). Birmingham City Council. . Retrieved 7 June 2008.
[123] "The Smethwick Engine" (http://www.matthewboulton2009.org/index.php/gallery/33-the-smethwick-engine). The Boulton 2009 Partnership. . Retrieved 19 September 2010.
[124] "Nightlife in the City Centre" (http://web.archive.org/web/20080610221029/http://www.birmingham.gov.uk/GenerateContent?CONTENT_ITEM_ID=24761&CONTENT_ITEM_TYPE=0&MENU_ID=12535). Birmingham City Council. Archived from the original (http://www.birmingham.gov.uk/GenerateContent?CONTENT_ITEM_ID=24761&CONTENT_ITEM_TYPE=0&MENU_ID=12535) on 10 June 2008. . Retrieved 7 June 2008.
[125] "Birmingham's New Leisure Complex" (http://web.archive.org/web/20071212014801/http://www.memonline.com/news8.html). MEM Online News. Archived from the original (http://www.memonline.com/news8.html) on 12 December 2007. . Retrieved 7 June 2008.
[126] "Will a fest by any other name smell as sweet?" (http://www.birminghammail.net/birmingham-blogs-views/birmingham-mail-columnists/the-stirrer/tm_method=full&objectid=18192354&siteid=50002-name_page.html). Birmingham Mail. 1 December 2006. . Retrieved 7 June 2008.
[127] "History of St. Patrick's Day" (http://www.anysubject.com/st-patricks-day-history-of-st-patricks-day.asp). AnySubject. . Retrieved 7 June 2008.
[128] "Christmas markets" (http://www.enjoyengland.com/attractions/events/calendar/december/christmas-markets.aspx). Enjoy England. . Retrieved 11 October 2010.
[129] "Another Record year for Frankfurt Christmas Market" (http://www.birminghamnewsroom.com/?p=6019). Birmingham City Council. . Retrieved 29 September 2010.
[130] "Schedule of Nationally Listed Buildings of Historic Interest in Birmingham" (http://www.birmingham.gov.uk/ELibrary?E_LIBRARY_ID=75&a=1083070309470) (PDF). Birmingham City Council Planning Department. . Retrieved 7 June 2008.
[131] "The Lad In The Lane, Erdington" (http://web.archive.org/web/20071222043422/http://www.pub-explorer.com/wmids/pub/ladinlaneerdington.htm). pub-explorer.com. Archived from the original (http://www.pub-explorer.com/wmids/pub/ladinlaneerdington.htm) on 22 December 2007. .
[132] "History of Kings Norton" (http://web.archive.org/web/20080521222034/http://www.birmingham.gov.uk/GenerateContent?CONTENT_ITEM_ID=2474&CONTENT_ITEM_TYPE=0&MENU_ID=260). Birmingham City Council. Archived from the original (http://www.birmingham.gov.uk/GenerateContent?CONTENT_ITEM_ID=2474&CONTENT_ITEM_TYPE=0&MENU_ID=260) on 21 May 2008. . Retrieved 7 June 2008.
[133] Anne Baltz Rodrick (2004). *Self-Help and Civic Culture: Citizenship in Victorian Birmingham*. Ashgate Publishing. ISBN 0-7546-3307-1.
[134] "Birmingham's hidden jewel" (http://www.bbc.co.uk/birmingham/content/articles/2007/04/05/st_chads_feature.shtml). BBC Birmingham. . Retrieved 7 June 2008.
[135] Phil Jones. "Tower Block Modernism vs. Urban Morphology: An analysis of Lee Bank, Birmingham" (http://web.archive.org/web/20080227084303/http://www.gees.bham.ac.uk/research/cpp/membersfiles/phil/leebankabstract.pdf) (PDF). Archived from the original (http://www.gees.bham.ac.uk/research/cpp/membersfiles/phil/leebankabstract.pdf) on 27 February 2008. . Retrieved 7 June 2008.
[136] "Aerial View of New Street Station 1963" (http://web.archive.org/web/20080511152530/http://www.birmingham.gov.uk/GenerateContent?CONTENT_ITEM_ID=89806&CONTENT_ITEM_TYPE=0&MENU_ID=10277). Birmingham City Council. Archived from the original (http://www.birmingham.gov.uk/GenerateContent?CONTENT_ITEM_ID=89806&CONTENT_ITEM_TYPE=0&MENU_ID=10277) on 11 May 2008. . Retrieved 7 June 2008.
[137] "Castle Vale" (http://web.archive.org/web/20071012100538/http://www.birmingham.gov.uk/GenerateContent?CONTENT_ITEM_ID=19750&CONTENT_ITEM_TYPE=0&MENU_ID=1453). Birmingham City Council. Archived from the original (http://www.birmingham.gov.uk/GenerateContent?CONTENT_ITEM_ID=19750&CONTENT_ITEM_TYPE=0&MENU_ID=1453) on 12 October 2007. . Retrieved 7 June 2008.
[138] "Awards" (http://www.future-systems.com/company/awards.html). Future Systems. . Retrieved 7 June 2008.
[139] "Town Hall, Birmingham" (http://www.thsh.co.uk/page/town-hall-birmingham/). Birmingham City Council. . Retrieved 21 June 2010.
[140] "Birmingham High Places document" (http://web.archive.org/web/20080609045359/http://www.birmingham.gov.uk/GenerateContent?CONTENT_ITEM_ID=13971&CONTENT_ITEM_TYPE=0&MENU_ID=1453). Birmingham City Council. Archived from the original (http://www.birmingham.gov.uk/GenerateContent?CONTENT_ITEM_ID=13971&CONTENT_ITEM_TYPE=0&MENU_ID=1453) on 9 June 2008. . Retrieved 7 June 2008.
[141] "Canals in Birmingham" (http://web.archive.org/web/20080609052603/http://www.birmingham.gov.uk/canals). Birmingham City Council. Archived from the original (http://www.birmingham.gov.uk/canals) on 9 June 2008. . Retrieved 7 June 2008.

[142] "Dad, are we nearly there yet?" (http://www.bbc.co.uk/legacies/heritage/england/birmingham/article_1.shtml). BBC. . Retrieved 7 June 2008.

[143] "UK Airport Statistics: 2009 -annual" (http://www.caa.co.uk/default.aspx?catid=80&pagetype=88&sglid=3&fld=2009Annual). Caa.co.uk. . Retrieved 12 April 2011.

[144] "Scheduled Timetable" (http://www.birminghamairport.co.uk/~/media/Files/Pdfs/global/timetables/scheduled-timetable.ashx). Birmingham Airport. 2010. . Retrieved 7 November 2010.

[145] "Bus Services" (http://web.archive.org/web/20080522122603/http://www.centro.org.uk/wwwroot/Bus/Bus intro.asp). Centro. Archived from the original (http://www.centro.org.uk/wwwroot/Bus/Bus intro.asp) on 22 May 2008. . Retrieved 7 June 2008.

[146] "The amazing number 11 bus" (http://www.bbc.co.uk/birmingham/content/articles/2005/06/27/number_11_bus_feature.shtml). BBC Birmingham. 27 June 2005. . Retrieved 7 June 2008.

[147] Barbara Elsom (21 June 2005). "Route 11 Bus Showcase" (http://www.birmingham.gov.uk/GenerateContent?CONTENT_ITEM_ID=61780&CONTENT_ITEM_TYPE=9&MENU_ID=276&CONTENT_TITLE=Route 11 Bus Showcase). Birmingham City Council. . Retrieved 7 June 2008. "Since 2001 231 bus stops out of 272 have been upgraded to Showcase standards..."

[148] Delta Rail (February 2011). "Station Usage 2009/10" (http://www.rail-reg.gov.uk/upload/pdf/stn_usage_report_0910.pdf) (PDF). Office of Rail Regulation. p. 6. . Retrieved 12 February 2011.; "Footfall Figures" (http://www.networkrail.co.uk/documents/5318_Footfall figures for 2007.pdf) (PDF). Network Rail. p. 1. . Retrieved 12 February 2011.

[149] "Travelmetro.co.uk" (http://www.travelmetro.co.uk/). Centro. . Retrieved 7 June 2008.

[150] "The Midland Metro (Birmingham City Centre Extension, etc.) Order 2005" (http://www.opsi.gov.uk/si/si2005/20051794.htm). Office of Public Sector Information. 2005. . Retrieved 7 June 2008.

[151] "Birmingham City Council Primary and Secondary" (http://web.archive.org/web/20070928104515/http://www.bgfl.org/services/nqt/primsec.htm). Birmingham Grid for Learning (BGfL). Archived from the original (http://www.bgfl.org/services/nqt/primsec.htm) on 28 September 2007. . Retrieved 7 June 2008.

[152] "Birmingham City Council Special Needs Schools" (http://web.archive.org/web/20070928104556/http://www.bgfl.org/services/nqt/sen.htm). Birmingham Grid for Learning (BGfL). Archived from the original (http://www.bgfl.org/services/nqt/sen.htm) on 28 September 2007. . Retrieved 7 June 2008.

[153] "Facts about Birmingham Library Service" (http://web.archive.org/web/20080609080413/http://www.birmingham.gov.uk/GenerateContent?CONTENT_ITEM_ID=1787&CONTENT_ITEM_TYPE=0&MENU_ID=5255). Birmingham City Council. Archived from the original (http://www.birmingham.gov.uk/GenerateContent?CONTENT_ITEM_ID=1787&CONTENT_ITEM_TYPE=0&MENU_ID=5255) on 9 June 2008. . Retrieved 7 June 2008.

[154] "Birmingham Adult Education Service" (http://www.birmingham.gov.uk/adulteducation). Birmingham City Council. . Retrieved 7 June 2008.

[155] "Birmingham Mobile Library Service" (http://www.birmingham.gov.uk/mobilelibrary). Birmingham City Council. . Retrieved 7 June 2008.

[156] "Grammar Schools of King Edward the Sixth" (http://web.archive.org/web/20080213094554/http://www.kingedwardthesixth.org/grammar.htm). The Grammar Schools of King Edward the Sixth in Birmingham. Archived from the original (http://www.kingedwardthesixth.org/grammar.htm) on 13 February 2008. . Retrieved 7 June 2008.

[157] Naqvi, Shahid (7 September 2007). "City college wins right to grant degrees" (http://icbirmingham.icnetwork.co.uk/birminghampost/news/tm_headline=city-college-wins-right-to-grant-degrees&method=full&objectid=19750162&siteid=50002-name_page.html). Birmingham Post. . Retrieved 7 June 2008.

[158] "A New Year and a new name for the College." (http://www.bcftcs.ac.uk/asp/UCB-faq.asp). University College Birmingham. . Retrieved 7 January 2008.

[159] "Sutton Coldfield College" (http://www.sutcol.ac.uk). . Retrieved 7 June 2008.

[160] "Joseph Chamberlain College" (http://web.archive.org/web/20080205191644/http://www.jcc.ac.uk/welcome.php). Archived from the original (http://www.jcc.ac.uk/welcome.php) on 5 February 2008. . Retrieved 7 June 2008.

[161] "KS07 Religion: Census 2001, Key Statistics for local Authorities" (http://www.statistics.gov.uk/statbase/ssdataset.asp?vlnk=6563&More=Y) (excel). *Census 2001*. Office for National Stistics. 13 February 2003. . Retrieved 12 September 2010.

[162] "Birmingham Central Mosque" (http://www.bbc.co.uk/birmingham/faith/places_of_worship/muslim/birmingham_central_mosque.shtml). BBC Birmingham Faith. . Retrieved 7 June 2008.

[163] "New city mosque a Symbol of Peace" (http://icbirmingham.icnetwork.co.uk/0100news/0100localnews/tm_objectid=14707141&method=full&siteid=50002&headline=new-city-mosque-a-symbol-of-peace-name_page.htmlicBirmingham). Icbirmingham.icnetwork.co.uk. 1 October 2004. . Retrieved 12 April 2011.

[164] "History of the Football League" (http://www.football-league.co.uk/page/History/HistoryDetail/0,,10794~1357277,00.html). The Football League. . Retrieved 30 December 2009.

[165] "Lawn Tennis and Major T. H. Gem" (http://www.birminghamcivicsociety.org.uk/lawntennis.htm). Birmingham Civic Society. . Retrieved 7 June 2008.

[166] Tyzack, Anna (22 June 2005). "The True Home of Tennis" (http://www.countrylife.co.uk/news/culture/article/79487/The_True_Home_of_Tennis.html). *Country Life* (IPC Media). . Retrieved 17 January 2009.

[167] Davis, Alex E (1988). *First in the field: the history of the world's first cricket league: the Birmingham and District Cricket League, formed 1888*. Brewin Books. ISBN 0947731342.

[168] "ICC Women's World Cup Qualifier schedule" (http://www.icc-europe.org/DATABASE/ARTICLES/articles/000035/003593.shtml). International Cricket Council. . Retrieved 14 September 2009.

[169] "Birmingham – We love our sport" (http://www.marketingbirmingham.com/media_office/fun_facts/fun_facts_sport/). Marketing Birmingham. . Retrieved 7 June 2008.

[170] Hill, Christopher R. (1994). "The Politics of Manchester's Olympic Bid" (http://pa.oxfordjournals.org/content/47/3/338.full.pdf+html). *Parliamentary Affairs* (The Hansard Society) **47** (3): 338–354. ISSN 0031-2290. . Retrieved 24 October 2010.

[171] Toohey, Kristine; James Veal, Anthony (2007). *The Olympic games: a social science perspective* (http://books.google.co.uk/books?id=ywy9aslk3M8C&pg=PA223). CABI. p. 223. ISBN 0851998097. . Retrieved 24 October 2010.

[172] The Second City derby (http://www.footballderbies.com/honours/index.php?id=27), footballderbies.com. Retrieved 30 December 2009.

[173] "Alumni – Brian Lara" (http://web.archive.org/web/20071204030409/http://www.thebears.co.uk/history/alumni/brianlara.shtml). Warwickshire County Cricket Club. Archived from the original (http://www.thebears.co.uk/history/alumni/brianlara.shtml) on 4 December 2007. . Retrieved 7 June 2008.

[174] "Contact the PGA" (http://www.pga.info/TheProfessionalGolfersAssociation/26363077.htm). The Professional Golfers Association. . Retrieved 2011-09-17.

[175] Bisset, Fergus. "England - Birmingham" (http://www.golf-monthly.co.uk/travel/uk/251753/england-birmingham.html). *Golf Monthly*. IPC Media. . Retrieved 2011-09-17.

[176] "Forest of Arden Country Club: Golf offerings and general information" (http://www.marriottgolf.com/leisure/golf-courses/forest-of-arden-country-club/default.jsp?loc=IM59*1-1LK67J). Marriott International. . Retrieved 2011-09-17.

[177] "2010 WTA Tour Tournament Calendar" (http://www.wtatour.com/page/Calendar/0,,12781,00.html). *Sony Ericsson WTA Tour – Official Site of Women's Professional Tennis*. 2010. . Retrieved 23 October 2010.

[178] McCarthy, Nick (3 June 2010). "Edgbaston Priory Tennis Club planning multi-million pound transformation" (http://www.birminghampost.net/news/west-midlands-news/2010/06/03/edgbaston-priory-tennis-club-planning-multi-million-pound-transformation-65233-26579605/). *Birmingham Post* (Trinity Mirror Midlands). . Retrieved 23 October 2010.

[179] Elkes, Neil (2011-07-14). "UK Athletics confirms move to Birmingham's Alexander Stadium" (http://www.birminghampost.net/news/west-midlands-news/2011/07/14/uk-athletics-confirms-move-to-birmingham-s-alexander-stadium-65233-29049011/). *Birmingham Post* (Trinity Mirror Midlands). . Retrieved 2011-09-18.

[180] "Samsung Diamond League Calendar of Events" (http://www.diamondleague.com/en/Event-Calendar/). Samsung Diamond League. . Retrieved 2011-09-18.

[181] "Permit Events" (http://www.iaaf.org/WIM10/results/index.html). IAAF Permit Indoor Meetings. . Retrieved 2011-09-18.

[182] "Birmingham will not have Olympic swimming pool by 2012" (http://www.birminghammail.net/news/top-stories/2009/10/08/birmingham-will-not-have-olympic-swimming-pool-by-2012-97319-24881847/). Birmingham Mail. 8 October 2009. . Retrieved 8 January 2010.

[183] Dale, Paul (3 March 2009). "Birmingham Council set to give green light to Digbeth market scheme" (http://www.birminghampost.net/news/politics-news/2009/03/03/birmingham-council-set-to-give-green-light-to-digbeth-market-scheme-65233-23047367/). *Birmingham Post*. . Retrieved 29 October 2009.

[184] Mark, Shepherd (3 July 2009). "The Wholesale shebang: traders at Birmingham's Wholesale Market may have a new home at Prupim's Hub by 2012. But that will happen to the existing site?" (http://www.propertyweek.com/story.asp?sectioncode=39&storycode=3144234). *Property Week* (United Business Media). . Retrieved 30 October 2009.

[185] "Michelin stars 2009" (http://servicesv2.webmichelin.com/frontnews/servlet/GetElement?elementCode=54989). Michelin Group. . Retrieved 18 January 2009.

[186] "Birmingham Breweries" (http://www.midlandspubs.co.uk/breweries/birmingham.htm). Midlands Pubs.co.uk. . Retrieved 7 June 2008.

[187] "BID Broad Street" (http://web.archive.org/web/20070930105529/http://www.broadstreetbid.com/flash/index.html). Archived from the original (http://www.broadstreetbid.com/flash/index.html) on 30 September 2007. . Retrieved 7 June 2008.

[188] "Taste of the Orient sweet for Wing Yip" (http://www.birminghampost.net/tm_objectid=16631720&method=full&siteid=50002&headline=taste-of-the-orient-sweet-for-wing-yip-name_page.html). The Birmingham Post Midland Rich List 2006. 6 January 2006. . Retrieved 7 June 2008.

[189] "The Balti Experience" (http://www.birmingham.gov.uk/balti). Birmingham City Council. . Retrieved 19 December 2006.

[190] "The Electric Cinema website" (http://www.theelectric.co.uk/). . Retrieved 7 June 2008.

[191] Thom Gorst (1995). *The Buildings Around Us*. Taylor & Francis. p. 93. ISBN 0-419-19330-8.

[192] "Birmingham IMAX" (http://www.imax.ac/). Thinktank. . Retrieved 28 November 2007.

[193] Liam Kennedy (2004). *Remaking Birmingham: The Visual Culture of Urban Regeneration*. Routledge. p. 115. ISBN 0-415-28838-X.

[194] "About Us – Information about BBC English Regions" (http://www.bbc.co.uk/england/about.shtml). BBC. . Retrieved 7 June 2008.

[195] "Lights, campus, action for BBC Birmingham's Television Drama Village" (http://www.bbc.co.uk/pressoffice/pressreleases/stories/2005/05_may/09/drama_village.shtml). BBC Press Release. 9 May 2005. . Retrieved 7 June 2008.

[196] Lee Carey (1 February 2003). "Ever Decreasing Circles" (http://www.transdiffusion.org/emc/studioone/circles.php). Studio One. . Retrieved 10 May 2008.
[197] "Radio stations in the West Midlands" (http://www.radio-now.co.uk/central.htm#WM). Radio Now. . Retrieved 22 July 2009.
[198] "The Archers airs 15,000th episode" (http://news.bbc.co.uk/1/hi/entertainment/6124558.stm). BBC News. 7 November 2006. . Retrieved 28 November 2007.
[199] "Major John Hall-Edwards" (http://www.birmingham.gov.uk/xray). .
[200] "Facts about Birmingham". BirminghamNet.
[201] "Lunar Society" (http://www.birminghamuk.com/lunarsociety.htm). BirminghamUK. . Retrieved 7 June 2008.
[202] "International Links" (http://www.birmingham.gov.uk/cs/Satellite?c=Page&childpagename=European-Team/PageLayout&cid=1223092718469&pagename=BCC/Common/Wrapper/Wrapper). Birmingham City Council. . Retrieved 31 May 2010.
[203] "*Frankfurt -Partner Cities*" (http://web.archive.org/web/20071107080201/http://www.frankfurt.de/sixcms/detail.php?id=502645). 2008 Stadt Frankfurt am Main (http://www.frankfurt.de/). Archived from the original (http://www.frankfurt.de/sixcms/detail.php?id=502645) on 7 November 2007. . Retrieved 17 July 2009.
[204] "Brno – Partnerská města" (http://www.brno.cz/index.php?nav02=1985&nav01=34&nav03=1010&nav04=1016&nav05=1249&nav06=1272) (in Czech). 2006–2009 City of Brno (http://www.brno.cz/index.php?nav01=70&nav02=2224). . Retrieved 17 July 2009.
[205] "*Partner Cities of Lyon and Greater Lyon*" (http://www.lyon.fr/vdl/sections/en/villes_partenaires/villes_partenaires_2/?aIndex=1). 2008 Mairie de Lyon. . Retrieved 17 July 2009.
[206] "*Milano – Città Gemellate*" (http://www.comune.milano.it/portale/wps/portal/CDM?WCM_GLOBAL_CONTEXT=/wps/wcm/connect/ContentLibrary/In Comune/In Comune/Citt Gemellate). 2008 Municipality of Milan (Comune di Milano). . Retrieved 17 July 2009.
[207] "Sister Cities of Guangzhou" (http://www.gzwaishi.gov.cn/Item/3970.aspx). Guangzhou Foreign Affairs Office. . Retrieved 10 February 2010.
[208] "Partner Cities" (http://www.birmingham.gov.uk/cs/Satellite/twins?packedargs=website=4&rendermode=live). Birmingham City Council. . Retrieved 31 May 2010.
[209] "Birmingham, Alabama" (http://www.birminghamnet.com/birmingham.htm). BirminghamNet. . Retrieved 7 June 2008.
[210] http://www.gutenberg.org/etext/13926

External links

- Birmingham City Council (http://www.birmingham.gov.uk/)
- Local Strategic Partnership for Birmingham (http://www.bebirmingham.org.uk/)
- The Birmingham Civic Society (http://www.birminghamcivicsociety.org.uk/)
- The Birmingham Military Tattoo (http://www.birminghamtattoo.co.uk/)
- Birmingham Speedway (http://www.birminghambrummies.co/)
- Birmingham travel guide from Wikitravel

List of Acts of the Parliament of the United Kingdom, 1880–1899

UK Legislation
Acts of Parliament by states preceding the United Kingdom
Of the Kingdom of England
Before 1485 1485–1601 · 1603–1641 Interregnum (1642–1660) 1660–1699 · 1700–1706
Royal statutes, etc. issued before the development of Parliament
Of the Kingdom of Ireland
Before 1701 · 1701–1800
Of the Kingdom of Scotland
Up to 1707
Of the Kingdom of Great Britain
1707–1719 · 1720–1739 · 1740–1759 · 1760–1779 · 1780–1800
Acts of the Parliament of the United Kingdom
1801–1819 · 1820–1839 · 1840–1859 · 1860–1879 · 1880–1899 · 1900–1919 · 1920–1939 · 1940–1959 · 1960–1979 · 1980–1999 · 2000 to date
Delegated Legislation
Statutory Rules and Orders
United Kingdom 1894–1947 Northern Ireland 1922–1974
United Kingdom Statutory Instruments
1948 · 1949 · 1950 · 1951 · 1952 · 1953 · 1954 · 1955 · 1956 · 1957 · 1958 · 1959 · 1960 · 1961 · 1962 · 1963 · 1964 · 1965 · 1966 · 1967 · 1968 · 1969 · 1970 · 1971 · 1972 · 1973 · 1974 · 1975 · 1976 · 1977 · 1978 · 1979 · 1980 · 1981 · 1982 · 1983 · 1984 · 1985 · 1986 · 1987 · 1988 · 1989 · 1990 · 1991 · 1992 · 1993 · 1994 · 1995 · 1996 · 1997 · 1998 · 1999 · 2000 · 2001 · 2002 · 2003 · 2004 · 2005 · 2006 · 2007 · 2008 · 2009 · 2010 · 2011
Devolved Legislation
Acts
Acts of the Scottish Parliament (1999 to date) Acts of the N. Ireland Parliament (1921–1972) Acts of the Northern Ireland Assembly (2000–2002, 2007 to date) Acts of the Welsh Assembly (2011 to date)
Measures and Orders
Measures of the Welsh Assembly (2007-2011) Measures of the N. Ireland Assembly (1973-1974) Orders in Council for Northern Ireland

Delegated Legislation
Scottish Statutory Instruments
1999 · 2000 · 2001 · 2002 · 2003 · 2004 · 2005 · 2006 · 2007 · 2008 · 2009 · 2010 · 2011
Welsh Statutory Instruments
1999 · 2000 · 2001 · 2002 · 2003 · 2004 · 2005 · 2006 · 2007 · 2008 · 2009 · 2010 · 2011
Statutory Rules of Northern Ireland
1974 · 1975 · 1976 · 1977 · 1978 · 1979 · 1980 · 1981 · 1982 · 1983 · 1984 · 1985 · 1986 · 1987 · 1988 · 1989 · 1990 · 1991 · 1992 · 1993 · 1994 · 1995 · 1996 · 1997 · 1998 · 1999 · 2000 · 2001 · 2002 · 2003 · 2004 · 2005 · 2006 · 2007 · 2009 · 2010 · 2011
Church of England
Measures · Instruments

This is an *incomplete* **list of Acts of the Parliament of the United Kingdom for the years 1880–1899**. Note that the first parliament of the United Kingdom was held in 1801; parliaments between 1707 and 1800 were either parliaments of Great Britain or of Ireland). For Acts passed up until 1707 see List of Acts of the Parliament of England and List of Acts of the Parliament of Scotland. For Acts passed from 1707 to 1800 see List of Acts of the Parliament of Great Britain. See also the List of Acts of the Parliament of Ireland.

For Acts of the devolved parliaments and assemblies in the United Kingdom, see the List of Acts of the Scottish Parliament from 1999, the List of Acts of the Northern Ireland Assembly, and the List of Acts and Measures of the National Assembly for Wales; see also the List of Acts of the Parliament of Northern Ireland.

The number shown after each Act's title is its chapter number. Acts passed before 1963 are cited using this number, preceded by the year(s) of the reign during which the relevant parliamentary session was held; thus the *Union with Ireland Act 1800* is cited as "39 & 40 Geo. 3 c. 67", meaning the 67th Act passed during the session that started in the 39th year of the reign of George III and which finished in the 40th year of that reign. Note that the modern convention is to use Arabic numerals in citations (thus "41 Geo. 3" rather than "41 Geo. III"). Note also that Acts of the last session of the Parliament of Great Britain and the first session of the Parliament of the United Kingdom are both cited as "41 Geo. 3". Acts passed from 1963 onwards are simply cited by calendar year and chapter number.

All modern Acts have a short title, e.g. the *Local Government Act 2003*. Some earlier Acts also have a short title given to them by later Acts, such as by the *Short Titles Act 1896*.

1880–1889

1880

43 Vict.

- Appropriation Act 1880 c. 13
- Army Discipline and Regulation (Annual) Act 1880 c. 9
- Artizans and Labourers Dwellings Improvement (Scotland) Act 1880 c. 2
- Artizans' Dwellings Act (1868) Amendment Act (1879) Amendment Act 1880 c. 8
- Beer Dealers Retail Licences Act 1880 c. 6
- Companies Act 1880 c. 19
- Consolidated Fund (No. 1) Act 1880 c. 5
- Customs and Inland Revenue Act 1880 c. 14
- East India Loan (East Indian Railway Debentures) Act 1880 c. 10
- Exchequer Bills and Bonds Act 1880 c. 16
- Hypothec Abolition (Scotland) Act 1880 c. 12
- India Stock (Powers of Attorney) Act 1880 c. 11
- Indian Salaries and Allowances Act 1880 c. 3

- National Debt Act 1880 c. 15
- Parliamentary Elections and Corrupt Practices Act 1880 c. 18
- Relief of Distress (Ireland) Act 1880 c. 4
- Roads Amendment Act 1880 c. 7
- Seed Supply (Ireland) Act 1880 c. 1
- Town Councils and Local Boards Act 1880 c. 17

43 & 44 Vict.

- Annual Turnpike Acts Continuance Act 1880 c. 12
- Appropriation Act 1880 (Session 2) c. 40
- Bastardy Orders Act 1880 c. 32
- Births and Deaths Registration Act (Ireland), 1880 c. 13
- Burial Laws Amendment Act 1880 c. 41
- Census (Ireland) Act 1880 c. 28
- Census (Scotland) Act 1880 c. 38
- Census Act 1880 c. 37
- Consolidated Fund (No. 1) Act 1880 (Session 2) c. 3
- Consolidated Fund (No. 2) Act 1880 (Session 2) c. 30
- County Bridges Loans Extension Act 1880 c. 5
- County Court Jurisdiction in Lunacy (Ireland) Act 1880 c. 39
- Courts of Justice Building Amendment Act 1880 c. 29
- Criminal Law Amendment Act 1880 c. 45
- Debtors (Scotland) Act 1880 c. 34
- Drainage and Improvement of Lands (Ireland) Act 1880 c. 27
- Elementary Education Act 1880 c. 23
- Employers' Liability Act 1880 c. 42
- Exchequer Bills and Bonds Act 1880 (Session 2) c. 21
- Expiring Laws Continuance Act 1880 c. 48
- Glebe Loan (Ireland) Acts Amendment Act 1880 c. 2
- Great Seal Act 1880 c. 10
- Ground Game Act 1880 c. 47
- House Occupiers in Counties Disqualification Removal (Scotland) Act 1880 c. 6
- Industrial Schools Act Amendment Act 1880 c. 15
- Inland Revenue Act 1880 c. 20
- Irish Loans Act 1880 c. 44
- Isle of Man Loans Act 1880 c. 8
- Judicial Factors (Scotland) Act 1880 c. 4
- Married Women's Policies of Assurance (Scotland) Act 1880 c. 26
- Merchant Seamen (Payment of Wages and Rating) Act 1880 c. 16
- Merchant Shipping (Carriage of Grain) Act 1880 c. 43
- Merchant Shipping (Fees and Expenses) Act 1880 c. 22
- Merchant Shipping Act (1854) Amendment Act 1880 c. 18
- Metropolitan Board of Works (Money) Act 1880 c. 25
- Post Office (Money Orders) Act 1880 c. 33
- Public Works Loans Act 1880 c 1
- Railways Construction Amendment (Ireland) Act 1880 c. 31
- Relief of Distress (Ireland) Amendment Act 1880 c. 14

- Revenue Offices (Scotland) Holidays Act 1880 c. 17
- Savings Banks Act 1880 c. 36
- Spirits Act 1880 c. 24
- Statutes (Definition of Time) Act 1880 c. 9
- Taxes Management Act 1880 c. 19
- Union Assessment Act 1880 c. 7
- Universities and College Estates Amendment Act 1880 c. 46
- Universities of Oxford and Cambridge (Limited Tenures) Act 1880 c. 11
- Wild Birds Protection Act 1880 c. 35

1881 (44 & 45 Vict.)

- Alkali, &c. Works Regulation Act 1881 c. 37
- Annual Turnpike Acts Continuance Act 1881 c. 31
- Appropriation Act 1881 c. 56
- Army Act 1881 c. 58
- Army Discipline and Regulation (Annual) Act 1881 c. 9
- Bankruptcy and Cessio (Scotland) Act 1881 c. 22
- British Honduras (Court of Appeal) Act 1881 c. 36
- Burial Grounds (Scotland) Act 1855, Amendment Act 1881 c. 27
- Burial and Registration Acts (Doubts Removal) Act 1881 c. 2
- Central Criminal Court (Prisons) Act 1881 c. 64
- Consolidated Fund (No. 1) Act 1881 c. 1
- Consolidated Fund (No. 2) Act 1881 c. 8
- Consolidated Fund (No. 3) Act 1881 c. 15
- Consolidated Fund (No. 4) Act 1881 c. 50
- Conveyancing Act 1881 c. 41
- Coroners (Ireland) Act 1881 c. 35
- Corrupt Practices (Suspension of Elections) Act 1881 c. 42
- Court of Bankruptcy (Ireland) Officers and Clerks Act 1881 c. 23
- Customs (Officers) Act 1881 c. 30
- Customs and Inland Revenue Act 1881 c. 12
- East Indian Railway (Redemption of Annuities) Act 1881 c. 53
- Expiring Laws Continuance Act 1881 c. 70
- Fugitive Offenders Act 1881 c. 69
- Highways (Isle of Wight) Act 1881 c. 72
- Incumbents of Benefices Loans Extension Act 1881 c. 25
- India Office (Sale of Superfluous Land) Act 1881 c. 7
- India Office Auditor Act 1881 c. 63
- Indian Loan Act 1881 c. 54
- Inland Revenue Buildings Act 1881 c. 10
- Irish Church Act Amendment Act 1881 c. 71
- Judicial Committee Act 1881 c. 3
- Land Law (Ireland) Act 1881 c. 49
- Land Tax Commissioners (Names) Act 1881 c. 16
- Leases for Schools (Ireland) Act 1881 c. 65
- Local Government Board (Ireland) Amendment Act 1881 c. 28
- Local Taxation Returns (Scotland) Act 1881 c. 6

- Married Women's Property (Scotland) Act 1881 c. 21
- Metropolitan Board of Works (Money) Act 1881 c. 48
- Metropolitan Open Spaces Act 1881 c. 34
- Municipal Elections Amendment (Scotland) Act 1881 c. 13
- National Debt Act 1881 c. 55
- Newspaper Libel and Registration Act 1881 c. 60
- Patriotic Fund Act 1881 c. 46
- Peace Preservation (Ireland) Act 1881 c. 5
- Pedlars Act 1881 c. 45
- Petroleum (Hawkers) Act 1881 c. 67
- Petty Sessions Clerks (Ireland) Act 1881 c. 18
- Pollen Fishing (Ireland) Act 1881 c. 66
- Post Office (Land) Act 1881 c. 20
- Post Office (Newspaper) Act 1881 c. 19
- Presumption of Life Limitation (Scotland) Act 1881 c. 47
- Protection of Person and Property (Ireland) Act 1881 c. 4
- Public Loans (Ireland) Remission Act 1881 c. 32
- Public Works Loans Act 1881 c. 38
- Reformatory Institutions (Ireland) Act 1881 c. 29
- Regulation of the Forces Act 1881 c. 57
- Removal Terms (Burghs) (Scotland) Act 1881, c. 39
- Royal University of Ireland Act 1881 c. 52
- Sea Fisheries (Clam and Bait Beds) Act 1881 c. 11
- Solicitors Remuneration Act 1881 c. 44
- South Wales Bridges Act 1881 c. 14
- Statute Law Revision and Civil Procedure Act 1881 c. 59
- Stratified Ironstone Mines (Gunpowder) Act 1881 c. 26
- Summary Jurisdiction (Process) Act 1881 c. 24
- Summary Jurisdiction (Scotland) Act 1881 c. 33
- Sunday Closing (Wales) Act 1881 c. 61
- Superannuation Act 1881 c. 43
- Supreme Court of Judicature Act 1881 c. 68
- Tramways (Ireland) Amendment Act 1881 c. 17
- Universities Elections Amendment (Scotland) Act 1881 c. 40
- Veterinary Surgeons Act 1881 c. 62
- Wild Birds Protection Act 1881 c. 51

1882 (45 & 46 Vict.)

- Allotments Extension Act 1882 c. 80
- Ancient Monuments Protection Act 1882 c. 73
- Annual Turnpike Acts Continuance Act 1882 c. 52
- Appropriation Act 1882 c. 71
- Arklow Harbour Act 1882 c. 13
- Army (Annual) Act 1882 c. 7
- Arrears of Rent (Ireland) Act 1882 c. 47
- Artizans Dwellings Act 1882 c. 54
- Baths and Wash Houses Act 1882 c. 30

- Beer Dealers' Retail Licences (Amendment) Act 1882 c. 34
- Bills of Exchange Act 1882 c. 61
- Bills of Sale Act (1878) Amendment Act 1882 c. 43
- Boiler Explosions Act 1882 c. 22
- Bombay Civil Fund Act 1882 c. 45
- Casual Poor Act 1882 c. 36
- Citation Amendment (Scotland) Act 1882 c. 77
- Civil Imprisonment (Scotland) Act 1882 c. 42
- Commonable Rights Compensation Act 1882 c. 15
- Consolidated Fund (No. 1) Act 1882 c. 1
- Consolidated Fund (No. 2) Act 1882 c. 4
- Consolidated Fund (No. 3) Act 1882 c. 8
- Consolidated Fund (No. 4) Act 1882 c. 28
- Constabulary (Ireland) Amendment Act 1882 c. 63
- Conveyancing Act 1882 c. 39
- Copyright (Musical Compositions) Act 1882 c. 40
- Corn Returns Act 1882 c. 37
- Corrupt Practices (Suspension of Elections) Act 1882 c. 68
- County Court Amendment (Ireland) Act 1882 c. 29
- County Courts (Costs and Salaries) Act 1882 c. 57
- Customs and Inland Revenue Act 1882 c. 41
- Customs and Inland Revenue Buildings (Ireland) Act 1882 c. 17
- Divided Parishes and Poor Law Amendment Act 1882 c. 58
- Documentary Evidence Act 1882 c. 9
- Duke of Albany (Establishment) Act 1882 c. 5
- Educational Endowments (Scotland) Act 1882 c. 59
- Election of Representative Peers (Ireland) Act 1882 c. 26
- Electric Lighting Act 1882 c. 56
- Entail (Scotland) Act 1882 c. 53
- Expiring Laws Continuance Act 1882 c. 64
- Fishery Board (Scotland) Act 1882 c. 78
- Friendly Societies (Quinquennial Returns) Act 1882 c. 35
- General Police and Improvement (Scotland) Act 1882 c. 6
- Government Annuities Act 1882 c. 51
- Highway Rate Assessment and Expenditure Act 1882 c. 27
- India (Home Charges Arrears) Act 1882 c. 79
- Inferior Courts Judgments Extension Act 1882 c. 31
- Intermediate Education (Ireland) Act 1882 c. 69
- Interments (felo de se) Act 1882 c. 19
- Irish Reproductive Loan Fund Amendment Act 1882 c. 16
- Isle of Man (Officers) Act 1882 c. 46
- Labourers Cottages and Allotments (Ireland) Act 1882 c. 60
- Lunacy Regulation Amendment Act 1882 c. 82
- Married Women's Property Act 1882 c. 75
- Merchant Shipping (Colonial Inquiries) Act 1882 c. 76
- Merchant Shipping (Expenses) Act 1882 c. 55
- Metropolis Management and Building Acts (Amendment) Act 1882 c. 14

- Metropolitan Board of Works (Money) Act 1882 c. 33
- Military Manoeuvres Act 1882 c. 10
- Militia Act 1882 c. 49
- Militia Storehouses Act 1882 c. 12
- Municipal Corporations Act 1882 c. 50
- Passenger Vessels Licences Amendment (Scotland) Act 1882 c. 66
- Pensions Commutation Act 1882 c. 44
- Petty Sessions (Ireland) Act 1882 c. 24
- Places of Worship Sites Amendment Act 1882 c. 21
- Poor Rate Assessment and Collection Act 1869, Amendment Act 1882 c. 20
- Post Office (Parcels) Act 1882 c. 74
- Post Office (Reply Post Cards) Act 1882 c. 2
- Prevention of Crime (Ireland) Act 1882 c. 25
- Prison Charities Act 1882 c. 65
- Public Health (Fruit Pickers Lodgings) Act 1882 c. 23
- Public Health (Scotland) Act 1867, Amendment Act 1882 c. 11
- Public Offices Site Act 1882 c. 32
- Public Schools (Scotland) Teachers Act 1882 c. 18
- Public Works Loans Act 1882 c. 62
- Reserve Forces Act 1882 c. 48
- Revenue, Friendly Societies, and National Debt Act 1882 c. 72
- Settled Land Act 1882 c. 38
- Slate Mines (Gunpowder) Act 1882 c. 3
- Somersham Rectory Act 1882 c. 81
- South Wales Turnpike Roads Amendment Act 1882 c. 67
- Supreme Court of Judicature (Ireland) Act 1882 c. 70

1883 (46 & 47 Vict.)

- Agricultural Holdings (England) Act 1883 c. 61
- Agricultural Holdings (Scotland) Act 1883 c. 62
- Annual Turnpike Acts Continuance Act 1883 c. 21
- Appropriation Act 1883 c. 50
- Army (Annual) Act 1883 c. 6
- Bankruptcy Act 1883 c. 52
- Bills of Sale (Ireland) Act (1879) Amendment Act 1883 c. 7
- Borough Constables Act 1883 c. 44
- Cheap Trains Act 1883 c. 34
- Cholera Hospitals (Ireland) Act 1883 c. 48
- City of London Parochial Charities Act 1883 c. 36
- Companies Act 1883 c. 28
- Companies (Colonial Registers) Act 1883 c. 30
- Consolidated Fund (No. 1) Act 1883 c. 2
- Consolidated Fund (No. 2) Act 1883 c. 5
- Consolidated Fund (No. 3) Act 1883 c. 13
- Consolidated Fund (No. 4) Act 1883 c. 23
- Consolidated Fund (Permanent Charges Redemption) Act 1883 c. 1
- Constabulary and Police (Ireland) Act 1883 c. 14

- Corrupt Practices (Suspension of Elections) Act 1883 c. 46
- Corrupt and Illegal Practices Prevention Act 1883 c. 51
- Counterfeit Medal Act 1883 c. 45
- Customs and Inland Revenue Act 1883 c. 10
- Diseases Prevention (Metropolis) Act 1883 c. 35
- Education (Scotland) Act 1883 c. 56
- Epidemic and other Diseases Prevention Act 1883 c. 59
- Expiring Laws Continuance Act 1883 c. 40
- Explosive Substances Act 1883 c. 3
- Factory and Workshop Act 1883 c. 53
- Glebe Loan (Ireland) Acts Amendment Act 1883 c. 8
- Greenwich Hospital Act 1883 c. 32
- Irish Reproductive Loan Fund Amendment Act 1883 c. 33
- Isle of Man Harbours Act 1883 c. 9
- Labourers (Ireland) Act 1883 c. 60
- Lands Clauses (Umpire) Act 1883 c. 15
- Lord Alcester's Grant Act 1883 c. 16
- Lord Wolseley's Grant Act 1883 c. 17
- Medical Act (1858) Amendment Act 1883 c. 19
- Merchant Shipping (Fishing Boats) Act 1883 c. 41
- Metropolitan Board of Works (Money) Act 1883 c. 27
- Municipal Corporations Act 1883 c. 18
- National Debt Act 1883 c. 54
- National Gallery (Loan) Act 1883 c. 4
- Patents, Designs, and Trade Marks Act 1883 c. 57
- Payment of Wages in Public-houses Prohibition Act 1883 c. 31
- Poor Law Conferences Act 1883 c. 11
- Post Office (Money Orders) Act 1883 c. 58
- Prevention of Crime (Ireland) Act 1882, Amendment (Audience of Solicitors) Act 1883 c. 12
- Prison Service (Ireland) Act 1883 c. 25
- Provident Nominations and Small Intestacies Act 1883 c. 47
- Public Health Act 1875, (Support of Sewers,) Amendment Act 1883 c. 37 (Public Health Act 1875)
- Public Works Loans Act 1883 c. 42
- Registry of Deeds Office (Ireland) Holidays Act 1883 c. 20
- Relief of Distressed Unions (Ireland) Act 1883 c. 24
- Revenue, Act 1883 c. 55
- Sea Fisheries Act 1883 c. 22
- Sea Fisheries (Ireland) Act 1883 c. 26
- Statute Law Revision Act 1883 c. 39
- Statute Law Revision and Civil Procedure Act 1883 c. 49
- Supreme Court of Judicature (Funds, &c.) Act 1883 c. 29
- Tramways and Public Companies (Ireland) Act 1883 c. 43
- Trial of Lunatics Act 1883 c. 38

1884

47 & 48 Vict.

- Annual Turnpike Acts Continuance Act 1884 c. 52
- Annuity (Sir H. Brand), 1884 c. 1
- Appropriation Act 1884 c. 73
- Army (Annual) Act 1884 c. 8
- Bankruptcy Appeals (County Courts) Act 1884 c. 9
- Bankruptcy Frauds and Disabilities (Scotland) Act 1884 c. 16
- Bishopric of Bristol Act 1884 c. 66
- Building Societies Act 1884 c. 41
- Canal Boats Act 1884 c. 75
- Chartered Companies, 1884 c. 56
- Cholera Hospitals (Ireland) Act 1884 c. 59
- Cholera, &c. Protection (Ireland) Act 1884 c. 69
- Colonial Attorneys Relief Amendment Act 1884 c. 24
- Colonial Prisoners Removal Act 1884 c. 31
- Consolidated Fund (No. 1) Act 1884 c. 4
- Consolidated Fund (No. 2) Act 1884 c. 15
- Contagious Diseases (Animals) Act 1884 c. 13
- Contagious Diseases (Animals) Transfer of Parts of Districts Act 1884 c. 47
- Corrupt Practices (Suspension of Elections) Act 1884 c. 78
- County of Dublin Jurors' and Voters' Revision Act 1884 c. 35
- Criminal Lunatics Act 1884 c. 64
- Customs and Inland Revenue Act 1884 c. 25
- Disused Burial Grounds Act 1884 c. 72
- Dublin Science and Art Museum Act 1884 c. 6
- Elections (Hours of Poll) Act 1884 c. 34
- Expiring Laws Continuance Act 1884 c. 53
- Fisheries (Oyster, Crab, and Lobster) Act (1877) Amendment Act 1884 c. 26
- Freshwater Fisheries Act 1884 c. 11
- Great Seal Act 1884 c. 30
- Greek Marriages Act 1884 c. 20
- Improvement of Lands (Ecclesiastical Benefices) Act 1884 c. 67
- Indian Marine Service Act 1884 c. 38
- Intestates Estates Act 1884 c. 71
- Isle of Man Harbours Act 1884 c. 7
- Licensing (Evidence) Act 1884 c. 29
- Loans for Schools and Training Colleges (Ireland) Act 1884 c. 22
- London Brokers' Relief Act 1884 c. 3
- Married Women's Property Act 1884 c. 14
- Matrimonial Causes Act 1884 c. 68
- Metropolitan Asylum Board (Borrowing Powers) Act 1884 c. 60
- Metropolitan Board of Works, (Money) Act 1884 c. 50
- Metropolitan Police Act 1884 c. 17
- Municipal Elections (Corrupt and Illegal Practices) Act 1884 c. 70
- National Debt Act 1884 c. 2

- National Debt (Conversion of Stock) Act 1884 c. 23
- National School Teachers Amendment (Ireland) Act 1884 c. 45
- Naval Discipline Act 1884 c. 39
- Naval Enlistment Act 1884 c. 46
- Naval Pensions Act 1884 c. 44
- New Parishes Acts and Church Building Acts Amendment Act 1884 c. 65
- Newcastle Chapter Act 1884 c. 33
- Oyster Cultivation (Ireland) Act 1884 c. 48
- Pensions and Yeomanry Pay Act 1884 c. 55
- Post Office (Protection) Act 1884 c. 76
- Prison Act 1884 c. 51
- Prisons (Ireland) Amendment Act 1884 c. 36
- Prosecution of Offences Act 1884 c. 58
- Public Health (Confirmation of Byelaws) Act 1884 c. 12
- Public Health (Ireland) Amendment Act 1884 c. 77
- Public Health (Officers) Act 1884 c. 74
- Public Libraries Act 1884 c. 37
- Public Works Loans Act 1884 c. 49
- Reformatory and Industrial Schools (Manx Children) Act 1884 c. 40
- Revenue Act 1884 c. 62
- Royal Military Asylum Chelsea (Transfer) Act 1884 c. 32
- Sea Fisheries Act 1884 c. 27
- Sea and Coast Fisheries Fund (Ireland) Act 1884 c. 21
- Settled Land Act 1884 c. 18
- Sheriff Court Houses (Scotland) Amendment Act 1884 c. 42
- Summary Jurisdiction Act 1884 c. 43
- Summary Jurisdiction over Children (Ireland) Act 1884 c. 19
- Superannuation Act 1884 c. 57
- Supreme Court of Judicature Act 1884 c. 61
- Tramways and Public Companies (Ireland) Amendment Act 1884 c. 28
- Trustee Churches (Ireland) Act 1884 c. 10
- Trusts (Scotland) Amendment Act 1884 c. 63
- Valuation (Metropolis) Amendment Act 1884 c. 5
- Yorkshire Registries Act 1884 c. 54

48 & 49 Vict.

- Additional Income Tax Act 1884 c. 1
- Consolidated Fund (No. 1) Act 1884 (Sess. 2) c. 2
- Representation of the People Act 1884 c. 3 *(commonly known as the Third Reform Act)*
- Tramways and Public Companies (Ireland) Act 1883, Amendment Act 1884 c. 5
- Yorkshire Registries Amendment, 1884 c. 4

1885 (48 & 49 Vict.)

- Annual Turnpike Acts Continuance Act 1885 c. 37
- Annuity, Princess Beatrice, 1885 c. 24
- Appropriation Act 1885 c. 64
- Army (Annual) Act 1885 c. 8
- Artillery and Rifle Ranges Act 1885 c. 36
- Bankruptcy (Office Accommodation) Act 1885 c. 47
- Barristers Admission (Ireland) Act 1885 c. 20
- Burial Boards (Contested Elections) Act 1885 c. 21
- Cape of Good Hope (Advance) Act 1885 c. 7
- Cholera Hospitals (Ireland) Act 1885 c. 39
- Consolidated Fund (No. 2) Act 1885 c. 6
- Consolidated Fund (No. 3) Act 1885 c. 14
- Constabulary (Ireland) Redistribution Act 1885 c. 12
- County Officers and Courts (Ireland) Amendment Act 1885 c. 71
- Criminal Law Amendment Act 1885 c. 69
- Crown Lands Act 1885 c. 79
- Customs and Inland Revenue Act 1885 c. 51
- Earldom of Mar, 1885 c. 48
- East India Loan, 1885 c. 28
- East India Unclaimed Stock Act 1885 c. 25
- Ecclesiastical Commissioners Act 1840, Amendment Act 1885 c. 55
- Ecclesiastical Commissioners Act 1885 c. 31
- Educational Endowments (Ireland) Act 1885 c. 78
- Egyptian Loan Act 1885 c. 11
- Elections (Hours of Poll) Act 1885 c. 10
- Evidence by Commission Act 1885 c. 74
- Exchequer and Treasury Bills Act 1885 c. 44
- Expiring Laws Continuance Act 1885 c. 59
- Federal Council of Australasia Act 1885 c. 60
- Friendly Societies Amendment Act 1885 c. 27
- Greenwich Hospital Act 1885 c. 42
- Highway Act Amendment Act 1885 c. 13
- Honorary Freedom of Boroughs Act 1885 c. 29
- Housing of the Working Classes Act 1885 c. 72
- Indian Army Pension Deficiency Act 1885 c. 67
- Industrial Schools (Ireland) Act 1885 c. 19
- Labourers (Ireland) Act 1885 c. 77
- Local Loans Sinking Funds Act 1885 c. 30
- Lunacy Acts Amendment Act 1885 c. 52
- Medical Relief Disqualification Removal Act 1885 c. 46
- Metropolis Management Amendment Act 1885 c. 33
- Metropolitan Board of Works (Money) Act 1885 c. 50
- Metropolitan Police Staff Superannuation Act 1885 c. 68
- Metropolitan Streets Act 1885 c. 18
- Municipal Voters Relief Act 1885 c. 9
- National Debt Act 1885 c. 43

- Parliamentary Elections (Returning Officers) Act 1885 c. 62
- Parliamentary Elections Corrupt Practices Act 1885 c. 56
- Parliamentary Registration (Ireland) Act 1885 c. 17
- Patents, Designs, and Trade Marks (Amendment) Act 1885 c. 63
- Pluralities Acts Amendment Act 1885 c. 54
- Polehampton Estates Act 1885 c. 40
- Post Office (Sites) Act 1885 c. 45
- Prevention of Crimes Amendment Act 1885 c. 75
- Public Health (Members and Officers) Act 1885 c. 53
- Public Health (Ships, &c.) Act 1885 c. 35
- Public Health and Local Government Conferences Act 1885 c. 22
- Public Works Loans Act 1885 c. 65
- Purchase of Land (Ireland) Act 1885 c. 73
- Redistribution of Seats Act 1885 c. 23
- Registration Act 1885 c. 15
- Registration Amendment (Scotland) Act 1885 c. 16
- Registration Appeals (Ireland) Act 1885 c. 66
- Revising Barristers Act 1885 c. 57
- School Boards Act 1885 c. 38
- Sea Fisheries (Scotland) Amendment Act 1885 c. 70
- Secretary for Scotland Act 1885 c. 61
- Shannon Act 1885 c. 41
- Submarine Telegraph Act 1885 c. 49
- Telegraph Act 1885 c. 58
- Thames Preservation Act 1885 c. 76
- Tithe Rentcharge Redemption Act 1885 c. 32
- Union Officers (Ireland) Act 1885 c. 80
- Water Rate Definition Act 1885 c. 34
- Yorkshire Registries Amendment Act 1885 c. 26

1886

49 & 50 Vict.

- Appropriation Act 1886 c. 26
- Army (Annual) Act 1886 c. 8
- Bankruptcy (Agricultural Labourers' Wages) Act 1886 c. 28
- Bankruptcy (Office Accommodation) Act 1886 c. 12
- British North America Act 1886 c. 35 *(known in Canada as the Constitution Act, 1886)*
- Burial Grounds (Scotland) Amendment Act 1886 c. 21
- Burial of Drowned Persons Act 1886 c. 20
- Cape Race Lighthouse Act 1886 c. 13
- Coal Mines Act 1886 c. 40
- Companies Act 1886 c. 23
- Consolidated Fund (No. 1) Act 1886 c. 4
- Consolidated Fund (No. 2) Act 1886 c. 7
- Contagious Diseases (Animals) Act 1886 c. 32
- Contagious Diseases Acts Repeal Act 1886 c. 10

- Crofters' Holdings (Scotland) Act 1886 c. 29
- Customs Amendment Act 1886 c. 41
- Customs and Inland Revenue Act 1886 c. 18
- Drill Grounds Act 1886 c. 5
- Extraordinary Tithe Redemption Act 1886 c. 54
- Freshwater Fisheries Act 1886 c. 2
- Glebe Loan (Ireland) Acts Amendment Act 1886 c. 6
- Guardianship of Infants Act 1886 c. 27
- Idiots Act 1886 c. 25
- Incumbents of Benefices Loans Extension Act 1886 c. 34
- International Copyright Act 1886 c. 33
- Intoxicating Liquors (Sale to Children) Act 1886 c. 56
- Labourers (Ireland) Act 1886 c. 59
- Land Registry Act 1886 c. 1
- Land Tax Commissioners Names Act 1886 c. 47
- Lunacy (Vacating of Seats) Act 1886 c. 16
- Marriage Act 1886 c. 14
- Marriages Validity Act 1886 c. 3
- Married Women (Maintenance in case of Desertion) Act 1886 c. 52
- Medical Act 1886 c. 48
- Metropolitan Board of Works (Money) Act 1886 c. 44
- Metropolitan Police Act 1886 c. 22
- Metropolitan Police (Compensation) Act 1886 c. 11
- National Debt Act 1886 c. 19
- Oxford University (Justices) Act 1886 c. 31
- Parliamentary Elections (Returning Officers) Act (1875) Amendment Act 1886 c. 57
- Patents Act 1886 c. 37
- Patriotic Fund Act 1886 c. 30
- Peace Preservation (Ireland) Continuance Act 1886 c. 24
- Peterhead Harbour of Refuge Act 1886 c. 49
- Poor Law Loans and Relief (Scotland) Act 1886 c. 51
- Poor Relief (Ireland) Act 1886 c. 17
- Prison (Officers' Superannuation) Act 1886 c. 9
- Public Works Loans Act 1886 c. 45
- Public Works Loans (Ireland) Act 1886 c. 46
- Removal Terms (Scotland) Act 1886 c. 50
- Returning Officers (Scotland) Act 1886 c. 58
- Revising Barristers Act 1886 c. 42
- Revising Barristers (Ireland) Act 1886 c. 43
- Riot (Damages) Act 1886 c. 38
- Salmon and Freshwater Fisheries Act 1886 c. 39
- Sea Fishing Boats (Scotland) Act 1886 c. 53
- Shop Hours Regulation Act 1886 c. 55
- Sporting Lands Rating (Scotland) Act 1886 c. 15
- West Indian Incumbered Estates Act 1886 c. 36

50 Vict.

- Appropriation Act 1886, Session 2 c. 1
- Belfast Commission Act 1886 c. 4
- Expiring Laws Continuance Act 1886 c. 5
- Secret Service Money (Repeal) Act 1886 c. 2
- Submarine Telegraph Act 1886 c. 3

1887 (50 & 51 Vict.)

- Allotments Act 1887 c. 48
- Allotments and Cottage Gardens Compensation for Crops Act 1887 c. 26
- Appellate Jurisdiction Act 1887 c. 70
- Appropriation Act 1887 c. 50
- Army (Annual) Act 1887 c. 2
- Bankruptcy (Discharge and Closure) Act 1887 c. 66
- British Settlements Act 1887 c. 54
- Charitable Trusts Act 1887 c. 49
- Coal Mines Regulation Act 1887 c. 58
- Consolidated Fund (No. 1) Act 1887 c. 1
- Consolidated Fund (No. 2) Act 1887 c. 14
- Conversion of India Stock Act 1887 c. 11
- Conveyancing (Scotland) Acts (1874 and 1879) Amendment Act 1887 c. 69
- Copyhold Act 1887 c. 73
- Coroners Act 1887 c. 71
- County Courts (Expenses) Act 1887 c. 3
- Criminal Law and Procedure (Ireland) Act 1887 c. 20
- Criminal Procedure (Scotland) Act 1887 c. 35
- Crofters Holdings (Scotland) Act 1887 c. 24
- Customs Consolidation Act 1876, Amendment Act 1887 c. 7
- Customs and Inland Revenue Act 1887 c. 15
- Deeds of Arrangement Act 1887 c. 57
- Duke of Connaught's Leave, 1887 c. 10
- Escheat (Procedure) Act 1887 c. 53
- Expiring Laws Continuance Act 1887 c. 63
- Friendly Societies Act 1887 c. 56
- Incumbents Resignation Act 1871, Amendment Act 1887 c. 23
- Incumbents of Benefices Loans Extension Act 1886, Amendment Act 1887 c. 8
- Isle of Man (Customs) Act 1887 c. 5
- Land Law (Ireland) Act 1887 c. 33
- Lieutenancy Clerks Allowances Act 1887 c. 36
- Local Authorities (Expenses) Act 1887 c. 72
- Local Government (Boundaries) Act 1887 c. 61
- London Parks and Works Act 1887 c. 34
- Lunacy Districts (Scotland) Act 1887 c. 39
- Margarine Act 1887 c. 29
- Markets and Fairs (Weighing of Cattle) Act 1887 c. 27
- Merchandise Marks Act 1887 c. 28
- Merchant Shipping (Fishing Boats) Act 1887 c. 4

- Merchant Shipping (Miscellaneous) Act 1887 c. 62
- Metropolis Management (Battersea and Westminster) Act 1887 c. 17
- Metropolitan Board of Works (Money) Act 1887 c. 31
- Metropolitan Police Act 1887 c. 45
- Military Tramways Act 1887 c. 65
- National Debt and Local Loans Act 1887 c. 16
- Open Spaces Act 1887 c. 32
- Pensions (Colonial Service) Act 1887 c. 13
- Pluralities Act 1887 c. 68
- Police Disabilities Removal Act 1887 c. 9
- Prison (Officers Superannuation, Scotland) Act 1887 c. 60
- Probation of First Offenders Act 1887 c. 25
- Public Libraries Acts Amendment Act 1887 c. 22
- Public Libraries Consolidation (Scotland) Act 1887 c. 42
- Public Works Loans Act 1887 c. 37
- Public-houses, Hours of Closing (Scotland) Act 1887 c. 38
- Quarry (Fencing) Act 1887 c. 19
- Savings Banks Act 1887 c. 40
- Secretary for Scotland Act 1887 c. 52
- Settled Land Acts (Amendment) Act 1887 c. 30
- Sheriff of Lanarkshire, 1887 c. 41
- Sheriffs Act 1887 c. 55
- Stannaries Act 1887 c. 43
- Statute Law Revision Act 1887 c. 59
- Superannuation Act 1887 c. 67
- Supreme Court of Judicature (Ireland) Act 1887 c. 6
- Technical Schools (Scotland) Act 1887 c. 64
- Trinidad and Tobago Act 1887 c. 44
- Truck Amendment Act 1887 c. 46
- Truro Bishopric and Chapter Acts Amendment Act 1887 c. 12
- Trustee Savings Banks Act 1887 c. 47
- Trusts (Scotland) Act 1867, Amendment Act 1887 c. 18
- Valuation of Lands (Scotland) Amendment Act 1887 c. 51
- Water Companies (Regulation of Powers) Act 1887 c. 21

1888 (51 & 52 Vict.)

- Appropriation Act 1888 c. 61
- Army (Annual) Act 1888 c. 4
- Bail (Scotland) Act 1888 c. 36
- Borough Funds (Ireland) Act 1888 c. 53
- Companies Clauses Consolidation Act 1888 c. 48
- Consolidated Fund (No. 1) Act 1888 c. 1
- Consolidated Fund (No. 2) Act 1888 c. 16
- Consolidated Fund (No. 3) Act 1888 c. 26
- Copyright (Musical Compositions) Act 1888 c. 17
- County Courts Act 1888 c. 43
- County Electors Act 1888 c. 10

- Crofters Commission (Delegation of Powers) Act 1888 c. 63
- Customs (Wine Duty) Act 1888 c. 14
- Customs and Inland Revenue Act 1888 c. 8
- Electric Lighting Act 1888; c. 12
- Employers Liability Act 1880, Continued, 1888 c. 58
- Expiring Laws Continuance Act 1888 c. 38
- Factory and Workshop Amendment (Scotland) Act 1888 c. 22
- Fishery (Ireland) Act 1888 c. 30
- Friendly Societies Act 1888 c. 66
- Glebe Lands Act 1888 c. 20
- Hawkers Act 1888 c. 33
- Imperial Defence Act 1888 c. 32
- Inebriates Act 1888 c. 19
- Isle of Man (Customs) Act 1888 c. 7
- Land Charges Registration and Searches Act 1888 c. 51
- Land Law (Ireland) Act 1888; c. 13
- Law of Distress Amendment Act 1888 c. 21
- Law of Distress and Small Debts (Ireland) Act 1888 c. 47
- Law of Libel Amendment Act 1888 c. 64
- Lloyd's Signal Stations Act 1888 c. 29
- Local Bankruptcy (Ireland) Act 1888 c. 44
- Local Government Act 1888 c. 41
- Marriages Validation Act 1888 c. 28
- Merchant Shipping (Life Saving Appliances) Act 1888 c. 24
- Metropolitan Board (Commission) Act 1888 c. 6
- Metropolitan Board of Works (Money) Act 1888 c. 40
- Mortmain and Charitable Uses Act 1888 c. 42
- Municipal Local Bills (Ireland) Act 1888 c. 34
- National Debt (Conversion) Act 1888 c. 2
- National Debt (Supplemental) Act 1888 c. 15
- National Defence Act 1888 c. 31
- North Sea Fisheries Act 1888 c. 18
- Oaths Act 1888 c. 46
- Oude and Rohilkund Railway Purchase Act 1888 c. 5
- Patents, Designs, and Trade Marks Act 1888 c. 50
- Preferential Payments in Bankruptcy Act 1888 c. 62
- Probate Duties (Scotland and Ireland) Act 1888 c. 60
- Public Health (Buildings in Streets) Act 1888 c. 52
- Public Works Loans Act 1888 c. 39
- Purchase of Land (Ireland) Amendment Act 1888 c. 49
- Railway and Canal Traffic Act 1888 c. 25
- Recorders, Magistrates, and Clerks of the Peace Act 1888 c. 23
- Roads and Bridges (Scotland) Act 1878, Amendment Act 1888 c. 9
- Sand-Grouse Protection Act 1888 c. 55
- Sea Fisheries Regulation Act 1888 c. 54
- Solicitors Act 1888 c. 65
- Special Commission Act 1888 c. 35

- Statute Law Revision Act 1888 c. 3
- Statute Law Revision (No. 2) Act 1888 c. 57
- Suffragans Nomination Act 1888 c. 56
- Supreme Court of Judicature (Ireland) Amendment Act 1888 c. 27
- Timber (Ireland) Act 1888 c. 37
- Trustee Act 1888 c. 59
- Victoria University Act 1888 c. 45
- Westminster Abbey Act 1888 c. 11

1889 (52 & 53 Vict.)

- Advertising Stations (Rating) Act 1889 c. 27
- Agricultural Holdings (Scotland) Act 1889 c. 20
- Appropriation Act 1889 c. 70
- Arbitration Act 1889 c. 49
- Army (Annual) Act 1889 c. 3
- Army and Navy Audit Act 1889 c. 31
- Assizes Relief Act 1889 c. 12
- Basutoland and British Bechuanaland Marriage Act 1889 c. 38
- Board of Agriculture Act 1889 c. 30
- Canada (Ontario Boundary) Act 1889 c. 28
- Clerks of Session (Scotland) Regulation Act 1889 c. 54
- Coinage Act 1889 c. 58
- Commissioners for Oaths Act 1889 c. 10
- Companies Clauses Consolidation Act 1889, c. 37
- Consolidated Fund (No. 1) Act 1889 c. 1
- Consolidated Fund (No. 2) Act 1889 c. 2
- Consolidated Fund (No. 3) Act 1889 c. 15
- Cotton Cloth Factories Act 1889 c. 62
- Council of India Reduction Act 1889 c. 65
- County Court Appeals (Ireland) Act 1889 c. 48
- Customs and Inland Revenue Act 1889 c. 7
- Expiring Laws Continuance Act 1889 c. 67
- Factors Act 1889 c. 45
- Friendly Societies Act 1889 c. 22
- General Police and Improvement (Scotland) Act 1862, Amendment Act 1889 c. 51
- Herring Fishery (Scotland) Act 1889 c. 23
- Indecent Advertisements Act 1889 c. 18
- Infectious Disease (Notification) Act 1889 c. 72
- Interpretation Act 1889 c. 63
- Judicial Factors (Scotland) Act 1889, c. 39
- Land Law (Ireland) Act 1888, Amendment Act 1889 c. 59
- Light Railways (Ireland) Act 1889 c. 66
- Local Government (Scotland) Act 1889 c. 50
- London Coal Duties Abolition Act 1889 c. 17
- London Council (Money) Act 1889 c. 61
- Lunacy Acts Amendment Act 1889 c. 41
- Master and Servant Act 1889 c. 24

- Merchant Shipping Act 1889 c. 46
- Merchant Shipping (Colours) Act 1889 c. 73
- Merchant Shipping (Pilotage) Act 1889 c. 68
- Merchant Shipping (Tonnage) Act 1889 c. 43
- National Debt Act 1889 c. 6
- National Debt Redemption Act 1889 c. 4
- National Portrait Gallery Act 1889 c. 25
- Naval Defence Act 1889 c. 8
- Official Secrets Act 1889 c. 52
- Palatine Court of Durham Act 1889 c. 47
- Parliamentary Grant (Caithness and Sutherland) Act 1889 c. 75
- Passengers Acts Amendment Act 1889, c. 29
- Paymaster General Act 1889 c. 53
- Poor Law Act 1889 c. 56
- Preferential Payments in Bankruptcy (Ireland) Act 1889 c. 60
- Prevention of Cruelty to, and Protection of, Children Act 1889 c. 44
- Prince of Wales's Children Act 1889 c. 35
- Public Bodies Corrupt Practices Act 1889 c. 69
- Public Health Act 1889 c. 64
- Public Libraries Acts Amendment Act 1889, c. 9
- Public Works Loans Act 1889 c. 71
- Purchase of Land (Ireland) Amendment Act 1889 c. 13
- Registration of County Electors (Extension of Time) Act 1889 c. 19
- Regulation of Railways Act 1889 c. 57
- Removal of Wrecks Act 1877, Amendment Act 1889 c. 5
- Revenue Act 1889 c. 42
- Sale of Horseflesh, &c. Regulation Act 1889 c. 11
- Secretary for Scotland Act 1889 c. 16
- Settled Land Act 1889 c. 36
- Small Debt Amendment (Scotland) Act 1889, c. 26
- Steam Trawling (Ireland) Act 1889 c. 74
- Technical Instruction Act 1889 c. 76
- Telegraph (Isle of Man) Act 1889 c. 34
- Town Police Clauses Act 1889, c. 14
- Trust Investment Act 1889 c. 32
- Universities (Scotland) Act 1889 c. 55
- Weights and Measures Act 1889 c. 21
- Welsh Intermediate Education Act 1889 c. 40
- Windward Islands Appeal Court Act 1889 c. 33

1890 – 1899

1890 (53 & 54 Vict.)

- Allotments Act 1890 c. 65
- Anglo-German Agreement Act 1890 c. 32
- Appropriation Act 1890 c. 72
- Army (Annual) Act 1890 c. 4
- Bankruptcy Act 1890 c. 71
- Barracks Act 1890 c. 25
- Bills of Sale Act 1890 c. 53
- Boiler Explosions Act 1890 c. 35
- Census (England and Wales) Act 1890 c. 61
- Census (Ireland) Act 1890 c. 46
- Census (Scotland) Act 1890 c. 38
- Chancery of Lancaster Act 1890 c. 23
- Colonial Courts of Admiralty Act 1890 c. 27
- Commissioners for Oaths Amendment Act 1890 c. 7
- Companies (Memorandum of Association) Act 1890 c. 62
- Companies (Winding up) Act 1890 c. 63
- Consolidated Fund (No. 1) Act 1890 c. 1
- Consolidated Fund (No. 2) Act 1890 c. 28
- Contagious Diseases (Animals) (Pleuro-pneumonia) Act 1890 c. 14
- County Councils Association Expenses Act 1890 c. 3
- Crown Office Act 1890 c. 2
- Customs Consolidation Act 1876, Amendment Act 1890 c. 56
- Customs and Inland Revenue Act 1890 c. 8
- Deeds of Arrangement Amendment Act 1890 c. 24
- Directors Liability Act 1890 c. 64
- Education Code (1890) Act 1890 c. 22
- Education of Blind and Deafmute Children (Scotland) Act 1890 c. 43
- Elections (Scotland) (Corrupt and Illegal Practices) Act 1890 c. 55
- Electric Lighting (Scotland) Act 1890 c. 13
- Expiring Laws Continuance Act 1890 c. 49
- Factors (Scotland) Act 1890 c. 40
- Foreign Jurisdiction Act 1890 c. 37
- Herring Fishery (Scotland) Act Amendment Act 1890 c. 10
- Infectious Disease (Prevention) Act 1890 c. 34
- Inland Revenue Regulation Act 1890 c. 21
- Intestates' Estates Act 1890 c. 29
- Local Taxation (Customs and Excise) Act 1890 c. 60
- London County Council (Money) Act 1890 c. 41
- Lunacy Act 1890 c. 5
- Marriage Act 1890 c. 47
- Merchant Shipping Act 1890 c. 9
- Metropolis Management Act 1862, Amendment Act 1890 c. 54
- Metropolis Management Amendment Act 1890 c. 66
- Municipal Elections (Scotland) Act 1890 c. 11

- Open Spaces Act 1890 c. 15
- Parliamentary Registration Expenses (Ireland) Act 1890 c. 58
- Partnership Act 1890 c. 39
- Pauper Lunatic Asylums, Ireland, Superannuation Act 1890 c. 31
- Pharmacy Act (Ireland), 1875 c. 48
- Police Act 1890 c. 45
- Police (Scotland) Act 1890 c. 67
- Poor Law Acts (Ireland) Amendment Act 1890 c. 30
- Public Health (Rating of Orchards) Act 1890 c. 17
- Public Health Acts c. 59
- Public Health Amendment (Scotland) Act 1890 c. 20
- Public Libraries Acts Amendment Act 1890 c. 68
- Public Works Loans Act 1890 c. 50
- Railways (Ireland) Act 1890 c. 52
- Removal Terms (Scotland) Act 1886, Amendment Act 1890 c. 36
- Reserve Forces Act 1890 c. 42
- River Suck Drainage (Provision of Funds) Act 1890 c. 12
- Seed Potatoes Supply (Ireland) Art, 1890 c. 1
- Settled Land Act 1890 c. 69
- South Indian Railway Purchase Act 1890 c. 6
- Statute Law Revision (No. 2) Act 1890 c. 51
- Statute Law Revision Act 1890 c. 33
- Superannuation (War Department) Act 1890 c. 18
- Supreme Court of Judicature Act 1890 c. 44
- Tenants Compensation Act 1890 c. 57
- Transfer of Railways (Ireland) Act 1890 c. 2
- Trustees Appointment Act 1890 c. 19
- Western Australia Constitution Act 1890 c. 26
- Working Classes Act 1890 c. 70
- Working Classes Dwellings Act 1890 c. 16

1891 (54 & 55 Vict.)

- Allotments Rating Exemption Act 1891 c. 33
- Appropriation Act 1891 c. 55
- Army (Annual) Act 1891 c. 5
- Army Schools Act 1891 c. 16
- Bills of Sale Act 1891 c. 35
- Branding of Herrings (Northumberland) Act 1891 c. 28
- Brine Pumping (Compensation for Subsidence) Act 1891 c. 40
- Charitable Trusts (Recovery) Act 1891 c. 17
- Coinage Act 1891 c. 72
- Commissioners for Oaths Act 1891 c. 50
- Consolidated Fund (No. 1) Act 1891 c. 6
- Consolidated Fund (No. 2) Act 1891 c. 27
- Consular Salaries and Fees Act 1891 c. 36
- County Councils (Elections) Act 1891 c. 68
- Crofters Common Grazings Regulation Act 1891, c. 41

- Custody of Children Act 1891 c. 3
- Customs and Inland Revenue Act 1891 c. 25
- Electoral Disabilities Removal Act 1891 c. 11
- Elementary Education Act 1891, c. 56
- Expiring Laws Continuance Act 1891 c. 60
- Factory and Workshop Act 1891 c. 75
- Fisheries Act 1891 c. 37
- Foreign Marriage Act 1891 c. 74
- Forged Transfers Act 1891 c. 43
- Highways and Bridges Act 1891 c. 63
- Labourers (Ireland) Act 1891 c. 71
- Land Registry (Middlesex Deeds) Act 1891 c. 64
- Law Agents and Notaries Public (Scotland) Act 1891 c. 30
- Local Authorities Loans (Scotland) Act 1891 c. 34
- Local Registration of Title (Ireland) Act 1891 c. 66 (*also known as the Registration of Title Act 1891 in Ireland*)
- London County Council (Money) Act 1891 c. 62
- Lunacy Act 1891, c. 65
- Mail Ships Act 1891 c. 31
- Markets and Fairs (Weighing of Cattle) Act 1891 c. 70
- Merchandise Marks Act 1891 c. 15
- Metalliferous Mines (Isle of Man) Act 1891 c. 47
- Middlesex Registry Act 1891 c. 10
- Mortmain and Charitable Uses Act 1891 c. 73
- Museums and Gymnasiums Act 1891 c. 22
- Penal Servitude Act 1891, c. 69
- Pollen Fisheries (Ireland) Act 1891 c. 20
- Post Office Act 1891 c. 46
- Presumption of Life Limitation (Scotland) Act 1891 c. 29
- Public Accounts and Charges Act 1891 c. 24
- Public Health (London) Act 1891 c. 76
- Public Health (Scotland) Amendment Act 1891 c. 52
- Public Works Loans Act 1891 c. 59
- Purchase of Land (Ireland) Act 1891 c. 48
- Railway and Canal Traffic (Provisional Orders) Amendment Act 1891 c. 12
- Ranges Act 1891 c. 54
- Redemption of Rent (Ireland) Act 1891 c. 57
- Reformatory and Industrial Schools Act 1891 c. 23
- Registration of Certain Writs (Scotland) Act 1891 c. 9
- Registration of Electors Act 1891 c. 18
- Returning Officers (Scotland) Act 1891 c. 49
- Roads and Streets in Police Burghs (Scotland) Act 1891 c. 32
- Russian Dutch Loan Act 1891 c. 26
- Savings Banks Act 1891 c. 21
- Schools for Science and Art Act 1891 c. 61
- Seal Fishery (Behring's Sea) Act 1891 c. 19
- Seed Potatoes Supply (Ireland) Act 1891 c. 7
- Slander of Women Act 1891 c. 51

- Stamp Act 1891 c. 39
- Stamp Duties Management Act 1891 c. 38
- Statute Law Revision Act 1891 c. 67
- Supreme Court of Judicature Act 1891 c. 53
- Supreme Court of Judicature (London Causes) Act 1891 c. 14
- Taxes (Regulation of Remuneration) Act 1891 c. 13
- Technical Instruction Act 1891 c. 4
- Tithe Act 1891 c. 8
- Tramways (Ireland) Amendment Act 1891 c. 42
- Trusts (Scotland) Amendment Act 1891, c. 44
- Turbary (Ireland) Act 1891 c. 45
- Western Highlands and Islands (Scotland) Works Act 1891 c. 58

1892 (55 & 56 Vict.)

- Accumulations Act 1892 c. 58
- Alkali, &c. Works Regulation Act 1892 c. 30
- Allotments (Scotland) Act 1892 c. 54
- Ancient Monuments Protection (Ireland) Act 1892 c. 46
- Appropriation Act 1892 c. 33
- Army (Annual) Act 1892 c. 2
- Bank Act 1892 c. 48
- Betting and Loans (Infants) Act 1892 c. 4
- Birmingham Corporation Water Act
- Boards of Management of Poor Law District Schools (Ireland) Act 1892 c. 41
- British Columbia (Loan) Act 1892 c. 52
- Burgh Police (Scotland) Act 1892 c. 55
- Charity Inquiries (Expenses) Act 1892 c. 15
- Clergy Discipline Act 1892 c. 32
- Colonial Probates Act 1892 c. 6
- Colonial Stock Act 1892 c. 35
- Consolidated Fund (No. 1) Act 1892 c. 3
- Consolidated Fund (No. 2) Act 1892 c. 20
- Contagious Diseases (Animals) Act 1892 c. 47
- Conveyancing and Law of Property Act 1892 c. 13
- Coroners Act 1892 c. 56
- Customs and Inland Revenue Act 1892 c. 16
- Drainage and Improvement of Land (Ireland) Act 1892 c. 65
- Education and Local Taxation Account (Scotland) Act 1892 c. 51
- Expiring Laws Continuance Act 1892 c. 60
- Foreign Marriage Act 1892 c. 23
- Forged Transfers Act 1892 c. 36
- Gaming Act 1892 c. 9
- Hares Preservation Act 1892 c. 8
- High Court of Justiciary (Scotland) Act 1892 c. 21
- Housing of the Working Classes Act 1890, Amendment (Scotland) Act 1892 c. 22
- Indian Councils Act 1892 c. 14
- Irish Education Act 1892 c. 42

- Isle of Man Customs Act 1892 c. 28
- Labourers (Ireland) Act 1892 c. 7
- Land Commissioners (Ireland) Salaries Act 1892 c. 45
- Mauritius Hurricane Loan Act 1892 c. 49
- Merchant Shipping Act 1892 c. 37
- Military Lands Act 1892 c. 43
- Millbank Prison Act 1892 c. 1
- Mortmain and Charitable Uses Act Amendment Act 1892 c. 11
- National Debt (Conversion of Exchequer Bonds) Act 1892 c. 26
- National Debt (Stockholders Relief) Act 1892 c. 39
- Naval Knights of Windsor (Dissolution) Act 1892 c. 34
- Parliamentary Deposits and Bonds Act 1892 c. 27
- Police Returns Act 1892 c. 38
- Poor Law (Ireland) Act 1892 c. 5
- Post Office Act 1892 c. 24
- Private Street Works Act 1892 c. 57
- Public Libraries Act 1892 c. 53
- Public Works Loans Act 1892 c. 61
- Railway and Canal Traffic Act 1892 c. 44
- Roads and Bridges (Scotland) Amendment Act 1892; c. 12
- Salmon and Freshwater Fisheries Act 1892 c. 50
- Sheriff Courts (Scotland) Extracts Act 1892 c. 17
- Shop Hours Act 1892 c. 62
- Short Titles Act 1892 c. 10
- Small Holdings Act 1892 c. 31
- Statute Law Revision Act 1892 c. 19
- Superannuation Act 1892 c. 40
- Taxes (Regulation of Remuneration) Amendment Act 1892 c. 25
- Technical Instruction Amendment (Scotland) Act 1892 c. 63
- Technical and Industrial Institutions Act 1892 c. 29
- Telegraph Act 1892 c. 59
- Weights and Measures (Purchase) Act 1892 c. 18
- Witnesses (Public Inquiries) Protection Act 1892 c. 64

1893 (56 & 57 Vict.)

- Appeal (Forma Pauperis) Act 1893 c. 22
- Appropriation Act 1893 c. 60
- Army (Annual) Act 1893 c. 4
- Barbed Wire Act 1893 c. 32
- Burgh Police (Scotland) Act 1893 c. 25
- Burghs Gas Supply (Scotland) Act 1893 c. 52
- Cholera Hospitals (Ireland) Act 1893 c. 13
- Coinage Act 1893 c. 1
- Companies (Winding-up) Act 1893 c. 58
- Congested Districts Board (Ireland) Act 1893 c. 35
- Consolidated Fund (No. 1) Act 1893 c. 3
- Consolidated Fund (No. 2) Act 1893 c. 16

- Consolidated Fund (No. 3) Act 1893 c. 28
- Consolidated Fund (No. 4) Act 1893 c. 46
- Contagious Diseases (Animals) Act 1893 c. 43
- Conveyance of Mails Act 1893 c. 38
- County Surveyors (Ireland) Act 1893 c. 49
- Customs and Inland Revenue Act 1893 c. 7
- Day Industrial Schools (Scotland) Act 1893 c. 12
- Duchy of Cornwall Management Act 1893 c. 20
- East India Loan Act 1893 c. 70
- Elementary Education (Blind and Deaf Children) Act 1893 c. 42
- Elementary Education (School Attendance) Act 1893 c. 51
- Expiring Laws Continuance Act 1893 c. 59
- Fertilisers and Feeding Stuffs Act 1893 c. 56
- Friendly Societies Act 1893 c. 30
- Housing of the Working Classes Act 1893 c. 33
- Improvement of Land (Scotland) Act 1893 c. 34
- Industrial and Provident Societies Act 1893 c. 39
- Irish Education Act 1893 c. 41
- Isolation Hospitals Act 1893 c. 68
- Land Tax Commissioners Names Act 1893 c. 27
- Law of Commons Amendment Act 1893 c. 57
- Law of Distress and Small Debts, (Ireland) Act 1893 c. 36
- Light Railways (Ireland) Act 1893 c. 50
- Liverpool Court of Passage Act 1893 c. 37
- Local Authorities Loans (Scotland) Act 1891, Amendment Act 1893 c. 8
- Madras and Bombay Armies Act 1893 c. 62
- Married Women's Property Act 1893 c. 63
- Metropolis Management (Plumstead and Hackney) Act 1893 c. 55
- Municipal Corporations Act 1893 c. 9
- National Debt Redemption Act 1893 c. 64
- Naval Defence Act 1893 c. 45
- North Sea Fisheries Act 1893 c. 17
- Police Act 1893 c. 10
- Police Disabilities Removal Act 1893 c. 6
- Prison (Officers' Superannuation) Act 1893 c. 26
- Public Authorities Protection Act 1893 c. 61
- Public Health (London) Act 1891, Amendment Act 1893 c. 47
- Public Libraries (Amendment) Act 1893 c. 11
- Public Works Loans Act 1893 c. 24
- Public Works Loans (No. 2) Act 1893 c. 40
- Public Works Loans (No. 3) Act 1893 c. 65
- Railway Regulation Act 1893 c. 29
- Reformatory Schools Act 1893 c. 48
- Reformatory Schools (Scotland) Act 1893 c. 15
- Regimental Debts Act 1893 c. 5
- Rivers Pollution Prevention Acts, 1893 c. 31
- Rules Publication Act 1893 c. 66

- Sale of Goods Act 1893 c. 71
- Savings Bank Act 1893 c. 69
- Seal Fishery (North Pacific) Act 1893 c. 23
- Sheriff Courts Consignations (Scotland) Act 1893 c. 44
- Shop Hours Act 1893 c. 67
- Statute Law Revision Act 1893 c. 14
- Statute Law Revision (No. 2) Act 1893 c. 54
- Trade Union (Provident Funds) Act 1893 c. 2
- Treasury Chest Fund Act 1893 c. 18
- Trustee Act 1893 c. 53
- Voluntary Conveyances Act 1893 c. 21
- Weights and Measures Act 1893 c. 19

1894

56 & 57 Vict.

- Colonial Acts Confirmation Act 1894 c. 72
- Local Government Act 1894 c. 73

57 & 58 Vict.

- Appropriation Act 1894 c. 59
- Arbitration (Scotland) Act 1894 c. 13
- Army (Annual) Act 1894 c. 3
- Behring Sea Award Act 1894 c. 2
- Bishopric of Bristol Amendment Act 1894 c. 21
- British Museum (Purchase of Land) Act 1894 c. 34
- Building Societies Act 1894 c. 47
- Burgh Police (Scotland) Act 1892, Amendment Act 1894 c. 18
- Charitable Trusts (Places of Religious Worship) Amendment Act 1894 c. 35
- Chimney Sweepers Act 1894 c. 51
- Coal Mines (Check Weigher) Act 1894 c. 52
- Colonial Officers (Leave of Absence) Act 1894 c. 17
- Commissioners of Works Act 1894 c. 23
- Congested Districts Board (Ireland) Act 1894 c. 50
- Consolidated Fund (No. 1) Act 1894 c. 1
- Consolidated Fund (No. 2) Act 1894 c. 7
- Consolidated Fund (No. 3) Act 1894 c. 29
- Copyhold Act 1894 c. 46
- County Councils Association (Scotland) Expenses Act 1894 c. 5
- Crown Lands Act 1894 c. 43
- Diseases of Animals Act 1894 c. 57
- Expiring Laws Continuance Act 1894 c. 48
- Finance Act 1894 c. 30
- Fishery Board (Scotland) Extension of Powers Act 1894 c. 14
- Four Courts Library Act 1894 c. 4
- Heritable Securities (Scotland) Act 1894 c. 44
- Housing of the Working Classes Act 1894 c. 55

- Indian Railways Act 1894 c. 12
- Industrial Schools Acts Amendment Act 1894 c. 33
- Industrial and Provident Societies Act 1894 c. 8
- Injured Animals Act 1894 c. 22
- Jurors (Ireland) Amendment Act 1894 c. 49
- Local Government (Scotland) Act 1894 c. 58
- Locomotive Threshing Engines Act 1894 c. 37
- London (Equalisation of Rates) Act 1894 c. 53
- Merchandise Marks (Prosecutions) Act 1894 c. 19
- Merchant Shipping Act 1894 c. 60
- Music and Dancing Licences (Middlesex) Act 1894 c. 15
- Nautical Assessors (Scotland) Act 1894 c. 40
- Notice of Accidents Act 1894 c. 28
- Outdoor Relief Friendly Societies Act 1894 c. 25
- Prevention of Cruelty to Children Act 1894 c. 41
- Prevention of Cruelty to Children (Amendment) Act 1894 c. 27
- Prize Courts Act 1894 c. 39
- Public Libraries (Ireland) Act 1894 c. 38
- Public Libraries (Scotland) Act 1894 c. 20
- Public Works Loans Act 1894 c. 11
- Quarries Act 1894 c. 42
- Quarter Sessions Act 1894 c. 6
- Railway and Canal Traffic Act 1894 c. 54
- Registration Acceleration Act 1894 c. 32
- Sea Fisheries (Shell Fish) Regulation Act 1894 c. 26
- Solicitors Act 1894 c. 9
- Statute Law Revision Act 1894 c. 56
- Supreme Court of Judicature (Procedure) Act 1894 c. 16
- Trustee Act 1893, Amendment Act 1894 c. 10
- Uniforms Act 1894 c. 45
- Valuation of Lands (Scotland) Acts Amendment Act 1894 c. 36
- Wild Birds Protection Act 1894 c. 24
- Zanzibar Indemnity Act 1894 c. 31

1895

58 & 59 Vict.

- Appropriation Act 1895 c. 31
- Army (Annual) Act 1895 c. 7
- Australian Colonies Duties Act 1895 c. 3
- Colonial Boundaries Act 1895 c. 34
- Consolidated Fund (No. 1) Act 1895 c. 4
- Consolidated Fund (No. 2) Act 1895 c. 15
- Convention of Royal Burghs (Scotland) Act 1879, Amendment Act 1895 c. 6
- Corrupt and Illegal Practices Prevention Act 1895 c. 40
- Court of Session Consignations (Scotland) Act 1895 c. 19
- Courts of Law Fees (Scotland) Act 1895 c. 14

- Cruelty to Animals (Scotland) Act 1895 c. 13
- Documentary Evidence Act 1895 c. 9
- Extradition Act 1895 c. 33
- Factory and Workshop Act 1895 c. 37
- False Alarms of Fire Act 1895 c. 28
- Fatal Accidents Inquiry (Scotland) Act 1895 c. 36
- Finance Act 1895 c. 16
- Fisheries Close Season (Ireland) Act 1895 c. 29
- Friendly Societies Act 1895 c. 26
- Grand Jury (Ireland) Act 1895 c. 8
- Industrial and Provident Societies (Amendment) Act 1895 c. 30
- Isle of Man (Customs) Act 1895 c. 38
- Judicial Committee Amendment Act 1895 c. 44
- Lands Clauses (Taxation of Costs) Act 1895 c. 11
- Lands Valuation (Scotland) Amendment Act 1895 c. 41
- Law of Distress Amendment Act 1895 c. 24
- Local Government (Scotland) Act 1894, Amendment Act 1895 c. 1
- Local Government (Stock Transfer) Act 1895 c. 32
- Market Gardeners' Compensation Act 1895 c. 27
- Metropolitan Police (Receiver) Act 1895 c. 12
- Mortgagees Legal Costs Act 1895 c. 25
- Mr. Speaker's Retirement Act 1895 c. 10
- Naturalization Act 1895 c. 43
- Naval Works Act 1895 c. 35
- Out-door Relief (Ireland) Act 1895 c. 22
- Post Office Amendment Act 1895 c. 18
- Reformatory and Industrial Schools (Channel Islands Children) Act 1895 c. 17
- Sea Fisheries Regulation (Scotland) Act 1895 c. 42
- Seal Fisheries (North Pacific) Act 1895 c. 21
- Seed Potatoes Supply (Ireland) Act 1895 c. 2
- Shop Hours Act 1895 c. 5
- Summary Jurisdiction (Married Women) Act 1895 c. 39
- Tramways (Ireland) Act 1895 c. 20
- Volunteer Act 1895 c. 23

59 Vict. Sess. 2

Public General Acts

- Appropriation Act 1895, Session 2 c. 6
- Canadian Speaker (Appointment of Deputy) Act 1895, Session 2 c. 3
- Expiring Laws Continuance Act 1895, Session 2 c. 1
- Public Offices (Acquisition of Site) Act 1895, Session 2 c. 5
- Public Works Loans Act 1895, Session 2 c. 2
- Purchase of Land (Ireland) Amendment Act 1895, Session 2 c. 4

Local Acts

- Military Lands Provisional Order Confirmation (No. 2) Act 1895 Session 2 c. xv

1896 (59 & 60 Vict.)

- Agricultural Rates Act 1896 c. 16
- Agricultural Rates, Congested Districts, and Burgh Land Tax Relief (Scotland) Act 1896 c. 37
- Appropriation Act 1896 c. 46
- Army (Annual) Act 1896 c. 2
- Baths and Washhouses Act 1896 c. 59
- Bishopric of Bristol Amendment Act 1896 c. 29
- Burglary Act 1896 c. 57
- Chairmen of District Councils Act 1896 c. 22
- Coal Mines Regulation Act 1896 c. 43
- Collecting Societies and Industrial Assurance Companies Act 1896 c. 26
- Conciliation Act 1896 c. 30
- Consolidated Fund (No. 1) Act 1896 c. 3
- Consolidated Fund (No. 2) Act 1896 c. 7
- Derelict Vessels (Report) Act 1896 c. 12
- Diseases of Animals Act 1896 c. 15
- Dispensary Committees (Ireland) Act 1896 c. 10
- Edinburgh General Register House Act 1896 c. 24
- Expiring Laws Continuance Act 1896 c. 39
- Finance Act 1896 c. 28
- Fisheries (Norfolk and Suffolk) Act 1896 c. 18
- Friendly Societies Act 1896 c. 25
- Glasgow Parliamentary Divisions, Act 1896 c. 17
- Housing, of the Working Classes Act 1890, Amendment (Scotland) Act 1896 c. 31
- Housing of the Working Classes (Ireland) Act 1896 c. 11
- Incumbents of Benefices Loans Extension Act 1896 c. 13
- Judicial Trustees Act 1896 c. 35
- Labourers (Ireland) Act 1896 c. 53
- Land Law (Ireland) Act 1896 c. 47
- Larceny Act 1896 c. 52
- Law Agents (Scotland) Act Amendment Act 1896 c. 49
- Life Assurance Companies (Payment into Court) Act 1896 c. 8
- Light Railways Act 1896 c. 48
- Liverpool Court of Passage Act 1896 c. 21
- Local Government (Determination of Differences) Act 1896 c. 9
- Local Government (Elections) Act 1896 c. 1
- Local Government (Elections) (No. 2) Act 1896 c. 4
- Local Taxation (Ireland) Estate Duty Act 1896 c. 41
- Locomotives on Highways Act 1896 c. 36
- London Cab Act 1896 c. 27
- Naval Works Act 1896 c. 6
- Orkney and Zetland Small Piers and Harbours Act 1896 c. 32
- Poor Law Guardians (Ireland) (Women) Act 1896 c. 5
- Poor Law Officers' Superannuation Act 1896 c. 50

- Public Health Act 1896 c. 19
- Public Health (Ireland) Act 1896 c. 54
- Public Health (Ports) Act 1896 c. 20
- Public Offices (Westminster Site) Act 1896 c. 23
- Public Works Loans Act 1896 c. 42
- Quarter Sessions (London) Act 1896 c. 55
- Railways (Ireland) Act 1896 c. 34
- Royal Naval Reserve Volunteer Act 1896 c. 33
- Short Titles Act 1896 c. 14
- Stannaries Court (Abolition) Act 1896 c. 45
- Telegraph (Money) Act 1896 c. 40
- Truck Act 1896 c. 44
- Uganda Railway Act 1896 c. 38
- Vexatious Actions Act 1896 c. 51
- West Highland Railway Guarantee Act 1896 c. 58
- Wild Birds Protection Act 1896 c. 56

1897 (60 & 61 Vict.)

- Appropriation Act 1897 c. 67
- Archdeaconry of Cornwall Act 1897 c. 9
- Archdeaconry of London (Additional Endowment) Act 1897 c. 45
- Army (Annual) Act 1897 c. 3
- Chaff-Cutting Machines (Accidents) Act 1897 c. 60
- Cleansing of Persons Act 1897 c. 31
- Congested Districts (Scotland) Act 1897 c. 53
- Consolidated Fund (No. 1) Act 1897 c. 4
- Constabulary (Ireland) Act 1897 c. 64
- Cotton Cloth Factories Act 1897 c. 58
- County Dublin Surveyors Act 1897 c. 2
- Dangerous Performances Act 1897 c. 52
- District Councils (Water Supply Facilities) Act 1897 c. 44
- East India Company's Officers Superannuation Act 1897 c. 10
- Edinburgh University (Transfer of Patronage) Act 1897 c. 13
- Education (Scotland) Act 1897 c. 62
- Elementary Education Act 1897 c. 16
- Expiring Laws Continuance Act 1897 c. 54
- Extraordinary Tithe Act 1897 c. 23
- Finance Act 1897 c. 24
- Foreign Prison Made Goods Act 1897 c. 63
- Infant Life Protection Act 1897 c. 57
- Isle of Man (Church Building and New Parishes) Act 1897 c. 33
- Juries Detention Act 1897 c. 18
- Land Transfer Act 1897 c. 65
- Licensing Amendment (Scotland) Act 1897 c. 50
- Local Government Act 1897 c. 1
- Local Government (Joint Committees) Act 1897 c. 40
- Market Gardeners Compensation (Scotland) Act 1897 c. 22

- Merchant Shipping Act 1897 c. 59
- Merchant Shipping (Exemption from Pilotage) Act 1897 c. 61
- Mersey Channels Act 1897 c. 21
- Metropolis Water Act 1897 c. 56
- Metropolitan Police (Borrowing Powers) Act 1897 c. 42
- Metropolitan Police Courts Act 1897 c. 26
- Metropolitan Police Courts (Holidays) Act 1897 c. 14
- Military Lands Act 1897 c. 6
- Military Manoeuvres Act 1897 c. 43
- Military Works Act 1897 c. 7
- Municipal Elections (Scotland) Act 1897 c. 34
- Naval Works Act 1897 c. 35
- Navy and Marines (Wills) Act 1897 c. 15
- Out-door Relief (Ireland) Act 1897 c. 36
- Parish Councils Casual Vacancies (Scotland) Act 1897 c. 49
- Patent Office (Extension) Act 1897 c. 25
- Police (Property) Act 1897 c. 30
- Poor Law Act 1897 c. 29
- Poor Law Officers Superannuation Act Amendment Act 1897 c. 28
- Post Office and Telegraph Act 1897 c. 41
- Preferential Payments in Bankruptcy Amendment Act 1897 c. 19
- Public Health (Scotland) Act 1897 c. 38
- Public Offices (Whitehall) Site Act 1897 c. 27
- Public Works Loans Act 1897 c. 51
- Quarter Sessions Jurors (Ireland) Act 1897 c. 20
- Railway Assessors (Scotland) Superannuation Act 1897 c. 12
- Regular and Elders' Widows' Funds Act 1897 c. 11
- School Board Conference Act 1897 c. 32
- Stipendiary Magistrates Jurisdiction (Scotland) Act 1897 c. 48
- Supreme Court of Judicature (Ireland) Act 1897 c. 17
- Supreme Court of Judicature (Ireland) (No. 2) Act 1897 c. 66
- Trusts (Scotland) Act 1897 c. 8
- Voluntary Schools Act 1897 c. 5
- Volunteer Act 1897 c. 47
- Weights and Measures (Metric System) Act 1897 c. 46
- Wicklow Harbour Advances Act 1897 c. 55
- Workmen's Compensation Act 1897 c. 37
- Yorkshire Coroners Act 1897 c. 39

1898 (61 & 62 Vict.)

- Appropriation Act 1898 c. 61
- Army (Annual) Act 1898 c. 1
- Bail Act 1898 c. 7
- Benefices Act 1898 c. 48
- Canals Protection (London) Act 1898 c. 16
- Circuit Clerks (Scotland) Act 1898 c. 40
- Companies Act 1898 c. 26
- Consolidated Fund (No. 1) Act 1898 c. 3
- Consolidated Fund (No. 2) Act 1898 c. 32
- Criminal Evidence Act 1898 c. 36
- East India Loan Act 1898 c. 13
- Elementary School Teachers (Superannuation) Act 1898 c. 57
- Ex-officio Justices of the Peace (Scotland) Act 1898 c. 20
- Expiring Laws Continuance Act 1898 c. 47
- Finance Act 1898 c. 10
- Greek Loan Act 1898 c. 4
- Greenwich Hospital Act 1898 c. 24
- Inebriates Act 1898 c. 60
- Isle of Man (Customs) Act 1898 c. 27
- Kingstown Township (Transfer of Harbour Roads) Act 1898 c. 52
- Libraries Offences Act 1898 c. 53
- Local Government (Ireland) Act 1898 c. 37
- Local Taxation Account (Scotland) Act 1898 c. 56
- Locomotives Act 1898 c. 29
- Marriage Act 1898 c. 58
- Merchant Shipping (Liability of Shipowners) Act 1898 c. 14
- Merchant Shipping (Mercantile Marine Fund) Act 1898 c. 44
- Metropolitan Commons Act 1898 c. 43
- Metropolitan Police Courts Act 1898 c. 31
- Metropolitan Poor Act 1898 c. 45
- Mussels, Periwinkles and Cockles (Ireland) Act 1898 c. 28
- Out-door Relief (Ireland) Act 1898 c. 51
- Parish Fire-Engines Act 1898 c. 38
- Pauper Children (Ireland) Act 1898 c. 30
- Pharmacy Acts Amendment Act 1898 c. 25
- Poor Law (Scotland) Act 1898 c. 21
- Poor Law Unions Association (Expenses) Act 1898 c. 19
- Post Office (Guarantee) Act 1898 c. 18
- Post Office Guarantee (No. 2) Act 1898 c. 59
- Public Buildings Expenses Act 1898 c. 5
- Public Record Office Act 1898 c. 12
- Public Works Loans Act 1898 c. 54
- Prison Act 1898 c. 41
- Registration (Ireland) Act 1898 c. 2
- Reserve Forces and Militia Act 1898 c. 9
- Revenue Act 1898 c. 46

- Rivers Pollution Prevention (Border Councils) Act 1898 c. 34
- Seed Supply and Potato Spraying (Ireland) Act 1898 c.50
- Sheriffs Tenure of Office (Scotland) Act 1898 c. 8
- Societies' Borrowing Powers Act 1898 c. 15
- Solicitors (Ireland) Act 1898 c. 17
- Special Juries Act 1898 c. 6
- Statute Law Revision Act 1898 c. 22
- Suffragan Bishops Act 1898 c. 11
- Telegraph (Money) Act 1898 c. 33
- Trusts (Scotland) Act 1898 c. 42
- Union of Benefices Act 1898 c. 23
- Universities and College Estates Act 1898 c. 55
- University of London Act 1898 c. 62
- Vaccination Act 1898 c. 49
- Vagrancy Act 1898 c. 39
- Vexatious Actions (Scotland) Act 1898 c. 35

1899

62 & 63 Vict.

Public General Acts

- Agriculture and Technical Instruction (Ireland) Act 1899 c. 50
- Anchors and Chain Cables Act 1899 c. 23
- Appropriation Act 1899 c. 49
- Army (Annual) Act 1899 c. 3
- Baths and Washhouses Act 1899 c. 29
- Board of Education Act 1899 c. 33
- Bodies Corporate (Joint Tenancy) Act 1899 c. 20
- Colonial Loans Act 1899 c. 36
- Commons Act 1899 c. 30
- Congested Districts Board (Ireland) Act 1899 c. 18
- Consolidated Fund (No. 1) Act 1899 c. 2
- Electric Lighting (Clauses) Act 1899 c. 19
- Elementary Education (Defective and Epileptic Children) Act 1899 c. 32
- Elementary Education (School Attendance) Act (1893) Amendment Act 1899 c. 13
- Expiring Laws Continuance Act 1899 c. 34
- Finance Act 1899 c. 9
- Fine or Imprisonment (Scotland and Ireland) Act 1899 c. 11
- Gordon Memorial College at Khartoum Act 1899 c. 16
- Improvement of Land Act 1899 c. 46
- Inebriates Act 1899 c. 35
- Infectious Disease (Notification) Extension Act 1899 c. 8
- Isle of Man (Customs) Act 1899 c. 39
- Land Tax Commissioners Names Act 1899 c. 25
- Lincolnshire Coroners Act 1899 c. 48
- London Government Act 1899 c. 14
- Manchester Canonries Act 1899 c. 28

- Marriages Validity Act 1899 c. 27
- Metropolis Management Acts Amendment (Byelaws) Act 1899 c. 15
- Metropolis Water Act 1899 c. 7
- Metropolitan Police Act 1899 c. 26
- Military Works Act 1899 c. 41
- Naval Works Act 1899 c. 42
- Parish Councillors (Tenure of Office) Act 1899 c. 10
- Partridge Shooting (Ireland) Act 1899 c. 1
- Patriotic Fund Act 1899 c. 45
- Poor Law Act 1899 c. 37
- Private Legislation Procedure (Scotland) Act 1899 c. 47
- Public Libraries (Scotland) Act 1899 c. 5
- Public Works Loans Act 1899 c. 31
- Reformatory Schools Act 1899 c. 12
- Reserve Forces Act 1899 c. 40
- Royal Niger Company Act 1899 c. 43
- Sale of Food and Drugs Act 1899 c. 51
- Seats for Shop Assistants Act 1899 c. 21
- Small Dwellings Acquisition Act 1899 c. 44
- Solicitors Act 1899 c. 4
- Summary Jurisdiction Act 1899 c. 22
- Supreme Court of Judicature Act 1899 c. 6
- Telegraph Act 1899 c. 38
- Tithe Rentcharge (Rates) Act 1899 c. 17
- University of London Act 1899 c. 24

Local Acts

- Local Government Board (Ireland) Provisional Order Confirmation (No. 4) Act c. 34

62 & 63 Vict. Sess. 2

- Appropriation Act 1899, Session 2 c. 1
- Second Session (Explanation) Act 1899 c. 3
- Treasury Bills Act 1899 c. 2

Birmingham Corporation Water Department

The **Birmingham Corporation Water Department** was responsible for the supply of water to Birmingham from 1876 to 1974. It was also known as **Birmingham Corporation Waterworks Department**.

Early History 1808 - 1876

The earliest formal supply of water was offered by the Birmingham Waterworks Company. In 1808 notice of a Bill was given in Parliament for a scheme to provide Birmingham with an organised supply of water. It was opposed, and in 1809 a meeting considered the proposal, appointed a committee, and rejected the idea of the waterworks. A second Bill was rejected in 1811. It was not until 1826 that Parliament granted powers constituting

> "the Company of Proprietors of the Birmingham Waterworks for the purpose of providing a sufficient and constant supply of good and wholesome water for domestic, manufacturing and other purposes"

Birmingham Waterworks Tower by John Henry Chamberlain

The area to be supplied used only the River Tame. The initial supply was intermittent. In 1849, the Corporation made agreement with the Company for a constant supply in certain districts. By 1853 a constant supply was universal.

The Grade II listed[1] Birmingham Waterworks Tower at Edgbaston Reservoir was designed by John Henry Chamberlain and built in 1870. The reservoir was first built to feed the canals, a hundred years previously.

By 1873 the total quantity supplied was around 3,000 million imperial (**unknown operator: u'strong'unknown operator: u',**'m^3) per annum.

Birmingham Corporation Water Department

Early developments

The Birmingham Corporation Act of 1875 empowered Birmingham Corporation to purchase the Birmingham Waterworks Company. Takeover was orchestrated by Joseph Chamberlain Mayor of Birmingham and completed on 1 January 1876 for the sum of £1,350,000 (£94123430 as of 2011)[2] . Chamberlain declared to a House of Commons Committee

> We have not the slightest intention of making profit...We shall get our profit indirectly in the comfort of the town and in the health of the inhabitants

The Rivers Bourne and Blythe, Plant's Brook and Perry Stream were used as sources. The Corporation dug six wells at Aston, Short Heath, King's Vale, Perry Barr, Selly Oak, and Longbridge. These provided 20 million imperial gallons (91000 m^3) per day. There were 14 reservoirs with a total capacity of 628.5 million imperial gallons (m^3).

By 1891 population growth in the city was outstripping supply.

Elan aqueduct

Foel Tower in the Elan Valley

The Birmingham Corporation Water Act passed in 1892 authorised the purchase of land in South Wales on the upper portion of the Rivers Elan and Claerwen. Three reservoirs on the Elan and three on the Claerwen were authorised, together with an aqueduct to carry the water to Birmingham. The engineer for the Elan aqueduct scheme was James Mansergh. Construction work started in 1893 and the Elan Valley Railway was built to aid construction[3] . King Edward VII and Queen Alexandra performed the official opening on 21 July 1904 although building works were not completed towards the end of 1906.

The Elan aqueduct discharged into two reservoirs at Frankley Reservoir at Frankley and Bartley Reservoir at Bartley Green.[4]

The capacity of the aqueduct was increased with work starting in 1919. By 1921, two 42-inch-diameter (1100 mm) mains from Wales delivered about 25 million imperial gallons (m^3) per day.

Later history

City of Birmingham Water Department sign at Perry Barr Reservoir

Birmingham Corporation Water Department existed until 1974 when, under the Water Act 1973, its responsibilities were transferred to the **Severn Trent Water Authority** which is now Severn Trent Water.

See also

- Edgbaston Waterworks
- Selly Oak Pumping Station
- Elan Valley Reservoirs

References

[1] Details from listed building database (217751) - Grade II - Birmingham Waterworks Tower (http://www.imagesofengland.org.uk/Details/Default.aspx?id=217751). *Images of England*. English Heritage.
[2] UK CPI inflation numbers based on data available from Lawrence H. Officer (2010) " What Were the UK Earnings and Prices Then? (http://www.measuringworth.org/ukearncpi/)" MeasuringWorth.
[3] Steam nostalgia: locomotive and railway preservation in Great Britain. Gerald Nabarro
[4] Proceedings of the Institution of Mechanical Engineers, The Institution of Mechanical Engineers. 1910

Elan Valley Reservoirs

The **Elan Valley Reservoirs** are a chain of man-made lakes and reservoirs in the Elan Valley in Powys, Mid Wales (also known as the "Welsh Lake District"), using the rivers Elan and Claerwen. The reservoirs are Claerwen, Craig Goch, Pen-y-Garreg, Garreg Ddu, and Caban Coch.

Separating dam at Garreg Ddu

The work carried carried out over a hundred years ago to build the Elan Valley dams and reservoirs was only part of the huge undertaking. Almost as impressive was the challenge of delivering the enormous quantities of water by gravity alone, across very hilly country and over many river valleys, to the new Frankley Reservoir on the outskirts of Birmingham in the West Midlands of England. This involved building the 73 mile long Elan aqueduct, down which the water travels at less than 2 miles per hour, taking one and a half days to get to Birmingham.

The Elan aqueduct drops only 52m over a length of 118 km – a gradient of 1:2300. It runs from the Elan Valley to Frankley in Birmingham. There are several signs of the aqueduct between these points, in the form of brick aqueducts, exposed pipelines and red brick valve houses on hillsides.[1]

Birmingham's tap water is noted for being exceptionally soft, with a low percentage of dissolved minerals.[2]

Description

There are four main dams and reservoirs (constructed 1893–1904 in Elan Valley, and 1946–1952 at Claerwen) with a potential total capacity of nearly 100,000 megalitres.[3] The dams and reservoirs are:

Caban Goch Dam, Summer 2009

- Caban Coch with Garreg Ddu – 35,530 megalitre capacity
- Pen-y-garreg – 6,055 megalitre capacity
- Craig Goch – 9,222 megalitre capacity
- Claerwen – 48,300 megalitre capacity.

In addition to the four main dams, there are three other dams at the site:

- The Dol y Mynach dam – the masonry foundations and the base of this dam were laid in the Claerwen Valley at the time that the Elan Valley dams were being constructed. It was to be one of a series of three dams which were to hold back the waters of the Afon Claerwen making even more water available to the city of Birmingham in future times. The project was never completed but the base was laid in advance because the water level of the Caban Coch would have submerged the site once the reservoir was full. The project became completely redundant when newer materials and superior engineering led to the construction of the Claerwen dam higher up the valley.
- The Nant-y-Gro dam – this small dam was constructed in the early stages of the project to supply water to the navvies village at the site; it was most famously used during the Second World War by Sir Barnes Wallis during trials of the explosive charges he intended to use in the bouncing bombs which later targeted the Ruhr dams.
- The Garreg Ddu dam – although often referred to as a viaduct, the archways of this structure are built on a submerged dam; this hidden dam maintains the level of the reservoir behind it in times of extreme drought, and guarantees water can be extracted at the Foel Tower at sufficient height to enter the gravity driven aqueduct to the Frankley reservoir in Birmingham.

Garreg Ddu submerged dam and Foel Tower.

History

The reservoirs were constructed between 1893 and 1904 by the City of Birmingham's Water Department to supply clean water to the Birmingham area, by gravity feed along an aqueduct with a gradient of 1 in 2,300.

The Pen-y-Garreg dam overflowing in 2004

Before the construction of the dams, the standard gauge Elan Valley Railway was built to all dam sites from a junction of the Mid Wales Railway, at Rhayader. The railway was also built along the dams themselves at varying heights, on wooden scaffolding supported by concrete parapets. The railway itself went as far as the never-completed Dolymynach dam (lower down the valley from the later Claerwen dam), which had to be built concurrently with the Elan dams as the Caban Coch reservoir would otherwise have flooded this construction site. However, this railway was never subsequently needed, and road transport only was used for construction of the Claerwen dam over 40 years later.

The navvies (construction workers) lived in a village of wooden huts, which had a guard to prevent the illegal importing of liquor. This later became the permanent Elan Village (see Elan Valley). When construction of the dams was complete, most of the navvies moved on to the Derwent Valley in Derbyshire.

The scheme was opened by King Edward VII and Queen Alexandra on 21 July 1904.

World War II

They played an important role in World War II when the 35 foot (11 m) high Nant-y-Gro dam was used by Barnes Wallis to test his idea of detonating explosives against a dam wall in order to breach it. These experiments culminated in the Dambusters breaching of the dams in the Ruhr Valley. The remains of the breached Nant-y-Gro dam can still be seen today in the same state as it was left in 1942. The dam is now partially obscured by trees, but its location is marked by an interpretative plaque. The Derwent Dam was also used by the Dambusters for practice, though it was not breached.

After the Dambuster Raid steel cables were placed across Caban Coch reservoir to prevent enemy seaplanes landing on the lake.

Aborted expansion scheme

In the early 1970s it was proposed that the Craig Goch reservoir should be substantially increased in size with a new and higher downstream dam together with another dam to the north-west, impounding water that would otherwise have flowed down the Ystwyth valley. This scheme would have created a huge reservoir dwarfing all others in Mid Wales, and flooding miles of wild upland valleys. The proposals were eventually abandoned in the face of reducing projections for industrial water demand and an increasing awareness of the environmental issues that such an expansion might create.

The reservoirs today

The reservoirs are now owned by Dŵr Cymru Welsh Water. Although the filtration works further down the valley is run by Severn Trent Water.

A scale model of the reservoir network, in the form of ornamental ponds, is in Cannon Hill Park in Birmingham.

In popular culture

- Elizabeth Clarke wrote *The Valley* (published by Faber & Faber, 1969) an account of Welsh hill farming life in the valley between the two main periods of construction. Many references are made to the dams, as well as an account of the later enquiry from the hill farmers' point of view.
- Francis Brett Young researched the Elan Valley as the basis for his novel *The House under the Water* [Heinemann, 1932] imagining how the scene may have appeared in 1887.

See also

- Reservoirs and dams in the United Kingdom

References

Notes

[1] http://www.geograph.org.uk/photo/619851

[2] "How do I know if the hardness of my water is detrimental to my plumbing?" (http://www.aqua-nouveau.co.uk/products-service-and-information/frequently-asked-questions/#q1). Aqua-Nouveau. 2009. . Retrieved 2009-02-05.

[3] http://www.elanvalley.org.uk

Bibliography

- (N/A), (1955). *City of Birmingham Waterworks: A Short History of the Development of the Undertaking, with a Description of the Existing Works and Sources of Supply*. Birmingham: City of Birmingham, Water Committee. (Note: Published July 1954, on the 50 year Jubilee of the Elan Supply Scheme).
- Judge, Colin W. [1987] (1997). *The Elan Valley Railway: The Railway of the Birmingham Corporation Waterworks*. (OL 71) Headington: The Oakwood Press. (1997: ISBN 0-85361-517-9).

External links

- About the reservoirs (http://www.elanvalley.org.uk/dams_and_reservoirs.html)
- Dambusters at Nant-y-Gro dam (http://www.thedambusters.org.uk/nantygro.html)
- Aerial Photo of Claerwen (http://www.gtj.org.uk/item.php?lang=en&id=12318&t=1)
- Elan Historic Landscape (http://www.cpat.org.uk/projects/longer/histland/elan/evintr.htm)
- Abandoned Communities ... The Elan Valley reservoirs (http://www.abandonedcommunities.co.uk/page34.html)

52°16′20″N 3°41′20″W

Gravitation

Gravitation keeps the planets in orbits around the Sun. (Not to scale)

Gravitation, or **gravity**, is a natural phenomenon by which physical bodies attract with a force proportional to their mass. In everyday life, gravitation is most familiar as the agent that gives weight to objects with mass and causes them to fall to the ground when dropped. Gravitation causes dispersed matter to coalesce, and coalesced matter to remain intact, thus accounting for the existence of the Earth, the Sun, and most of the macroscopic objects in the universe. Gravitation is responsible for keeping the Earth and the other planets in their orbits around the Sun; for keeping the Moon in its orbit around the Earth; for the formation of tides; for natural convection, by which fluid flow occurs under the influence of a density gradient and gravity; for heating the interiors of forming stars and planets to very high temperatures; and for various other phenomena observed on Earth.

Gravitation is one of the four fundamental interactions of nature, along with electromagnetism, and the nuclear strong force and weak force. Modern physics describes gravitation using the general theory of relativity by Einstein, in which it is a consequence of the curvature of spacetime governing the motion of inertial objects. The simpler Newton's law of universal gravitation provides an accurate approximation for most physical situations.

History of gravitational theory

Scientific revolution

Modern work on gravitational theory began with the work of Galileo Galilei in the late 16th and early 17th centuries. In his famous (though possibly apocryphal[1]) experiment dropping balls from the Tower of Pisa, and later with careful measurements of balls rolling down inclines, Galileo showed that gravitation accelerates all objects at the same rate. This was a major departure from Aristotle's belief that heavier objects accelerate faster.[2] Galileo correctly postulated air resistance as the reason that lighter objects may fall more slowly in an atmosphere. Galileo's work set the stage for the formulation of Newton's theory of gravity.

Newton's theory of gravitation

In 1687, English mathematician Sir Isaac Newton published *Principia*, which hypothesizes the inverse-square law of universal gravitation. In his own words, "I deduced that the forces which keep the planets in their orbs must [be] reciprocally as the squares of their distances from the centers about which they revolve: and thereby compared the force requisite to keep the Moon in her Orb with the force of gravity at the surface of the Earth; and found them answer pretty nearly."[3]

Newton's theory enjoyed its greatest success when it was used to predict the existence of Neptune based on motions of Uranus that could not be accounted for by the actions of the other planets. Calculations by both John Couch Adams and Urbain Le Verrier predicted the general position of the planet, and Le Verrier's calculations are what led Johann Gottfried Galle to the discovery of Neptune.

A discrepancy in Mercury's orbit pointed out flaws in Newton's theory. By the end of the 19th century, it was known that its orbit showed slight perturbations that could not be accounted for entirely under Newton's theory, but all searches for another perturbing body (such as a planet orbiting the Sun even closer than Mercury) had been fruitless. The issue was resolved in 1915 by Albert Einstein's new theory of general relativity, which accounted for the small discrepancy in Mercury's orbit.

Although Newton's theory has been superseded, most modern non-relativistic gravitational calculations are still made using Newton's theory because it is a much simpler theory to work with than general relativity, and gives sufficiently accurate results for most applications involving sufficiently small masses, speeds and energies.

Equivalence principle

The equivalence principle, explored by a succession of researchers including Galileo, Loránd Eötvös, and Einstein, expresses the idea that all objects fall in the same way. The simplest way to test the weak equivalence principle is to drop two objects of different masses or compositions in a vacuum, and see if they hit the ground at the same time. These experiments demonstrate that all objects fall at the same rate when friction (including air resistance) is negligible. More sophisticated tests use a torsion balance of a type invented by Eötvös. Satellite experiments are planned for more accurate experiments in space.[4]

Formulations of the equivalence principle include:

- The weak equivalence principle: *The trajectory of a point mass in a gravitational field depends only on its initial position and velocity, and is independent of its composition.*[5]
- The Einsteinian equivalence principle: *The outcome of any local non-gravitational experiment in a freely falling laboratory is independent of the velocity of the laboratory and its location in spacetime.*[6]
- The strong equivalence principle requiring both of the above.

The equivalence principle can be used to make physical deductions about the gravitational constant, the geometrical nature of gravity, the possibility of a fifth force, and the validity of concepts such as general relativity and Brans-Dicke theory.

General relativity

In general relativity, the effects of gravitation are ascribed to spacetime curvature instead of a force. The starting point for general relativity is the equivalence principle, which equates free fall with inertial motion, and describes free-falling inertial objects as being accelerated relative to non-inertial observers on the ground.[7] [8] In Newtonian physics, however, no such acceleration can occur unless at least one of the objects is being operated on by a force.

Einstein proposed that spacetime is curved by matter, and that free-falling objects are moving along locally straight paths in curved spacetime. These straight paths are called geodesics. Like Newton's first law of motion, Einstein's theory states that if a force is applied on an object, it would deviate from a geodesic. For instance, we are no longer following geodesics while standing because the mechanical resistance of the Earth exerts an upward force on us, and we are non-inertial on the ground as a result. This explains why moving along the geodesics in spacetime is considered inertial.

Einstein discovered the field equations of general relativity, which relate the presence of matter and the curvature of spacetime and are named after him. The Einstein field equations are a set of 10 simultaneous, non-linear, differential equations. The solutions of the field equations are the components of the metric tensor of spacetime. A metric tensor describes a geometry of spacetime. The geodesic paths for a spacetime are calculated from the metric tensor.

Notable solutions of the Einstein field equations include:

- The Schwarzschild solution, which describes spacetime surrounding a spherically symmetric non-rotating uncharged massive object. For compact enough objects, this solution generated a black hole with a central singularity. For radial distances from the center which are much greater than the Schwarzschild radius, the accelerations predicted by the Schwarzschild solution are practically identical to those predicted by Newton's theory of gravity.
- The Reissner-Nordström solution, in which the central object has an electrical charge. For charges with a geometrized length which are less than the geometrized length of the mass of the object, this solution produces black holes with two event horizons.
- The Kerr solution for rotating massive objects. This solution also produces black holes with multiple event horizons.
- The Kerr-Newman solution for charged, rotating massive objects. This solution also produces black holes with multiple event horizons.
- The cosmological Friedmann-Lemaitre-Robertson-Walker solution, which predicts the expansion of the universe.

The tests of general relativity included the following:[9]

- General relativity accounts for the anomalous perihelion precession of Mercury.[2]
- The prediction that time runs slower at lower potentials has been confirmed by the Pound–Rebka experiment, the Hafele–Keating experiment, and the GPS.
- The prediction of the deflection of light was first confirmed by Arthur Stanley Eddington from his observations during the Solar eclipse of May 29, 1919.[10] [11] Eddington measured starlight deflections twice those predicted by Newtonian corpuscular theory, in accordance with the predictions of general relativity. However his interpretation of the results was later disputed.[12] More recent tests using radio interferometric measurements of quasars passing behind the Sun have more accurately and consistently confirmed the deflection of light to the degree predicted by general relativity.[13] See also gravitational lens.
- The time delay of light passing close to a massive object was first identified by Irwin I. Shapiro in 1964 in interplanetary spacecraft signals.
- Gravitational radiation has been indirectly confirmed through studies of binary pulsars.
- Alexander Friedmann in 1922 found that Einstein equations have non-stationary solutions (even in the presence of the cosmological constant). In 1927 Georges Lemaître showed that static solutions of the Einstein equations, which are possible in the presence of the cosmological constant, are unstable, and therefore the static universe envisioned by Einstein could not exist. Later, in 1931, Einstein himself agreed with the results of Friedmann and

Lemaître. Thus general relativity predicted that the Universe had to be non-static—it had to either expand or contract. The expansion of the universe discovered by Edwin Hubble in 1929 confirmed this prediction.[14]

- The theory's prediction of frame dragging was consistent with the recent Gravity Probe B results. [15]
- General relativity predicts that the light should lose its energy when travelling away from the massive bodies. The group of Radek Wojtak of the Niels Bohr Institute at the University of Copenhagen collected the data from 8000 galaxy clusters and found that the light coming from the cluster centers tended to be redshifted compared to the cluster egdes, confirming the energy loss due to the gravity. [16]

Gravity and quantum mechanics

In the decades after the discovery of general relativity it was realized that general relativity is incompatible with quantum mechanics.[17] It is possible to describe gravity in the framework of quantum field theory like the other fundamental forces, such that the attractive force of gravity arises due to exchange of virtual gravitons, in the same way as the electromagnetic force arises from exchange of virtual photons.[18] [19] This reproduces general relativity in the classical limit. However, this approach fails at short distances of the order of the Planck length,[17] where a more complete theory of quantum gravity (or a new approach to quantum mechanics) is required. Many believe the complete theory to be string theory,[20] or more currently M-theory, and, on the other hand, it may be a background independent theory such as loop quantum gravity or causal dynamical triangulation.

Specifics

Earth's gravity

Every planetary body (including the Earth) is surrounded by its own gravitational field, which exerts an attractive force on all objects. Assuming a spherically symmetrical planet, the strength of this field at any given point is proportional to the planetary body's mass and inversely proportional to the square of the distance from the center of the body.

The strength of the gravitational field is numerically equal to the acceleration of objects under its influence, and its value at the Earth's surface, denoted g, is approximately expressed below as the standard average.

$g = 9.81\ \mathrm{m/s^2} = 32.2\ \mathrm{ft/s^2}$

This means that, ignoring air resistance, an object falling freely near the Earth's surface increases its velocity by 9.81 m/s (32.2 ft/s or 22 mph) for each second of its descent. Thus, an object starting from rest will attain a velocity of 9.81 m/s (32.2 ft/s) after one second, 19.6 m/s (64.4 ft/s) after two seconds, and so on, adding 9.81 m/s (32.2 ft/s) to each resulting velocity. Also, again ignoring air resistance, any and all objects, when dropped from the same height, will hit the ground at the same time.

According to Newton's 3rd Law, the Earth itself experiences a force equal in magnitude and opposite in direction to that which it exerts on a falling object. This means that the Earth also accelerates towards the object until they collide. Because the mass of the Earth is huge, however, the acceleration imparted to the Earth by this opposite force is negligible in comparison to the object's. If the object doesn't bounce after it has collided with the Earth, each of them then exerts a repulsive contact force on the other which effectively balances the attractive force of gravity and prevents further acceleration.

Equations for a falling body near the surface of the Earth

Under an assumption of constant gravity, Newton's law of universal gravitation simplifies to $F = mg$, where m is the mass of the body and g is a constant vector with an average magnitude of 9.81 m/s^2. The acceleration due to gravity is equal to this g. An initially stationary object which is allowed to fall freely under gravity drops a distance which is proportional to the square of the elapsed time. The image on the right, spanning half a second, was captured with a stroboscopic flash at 20 flashes per second. During the first $^1/_{20}$ of a second the ball drops one unit of distance (here, a unit is about 12 mm); by $^2/_{20}$ it has dropped at total of 4 units; by $^3/_{20}$, 9 units and so on.

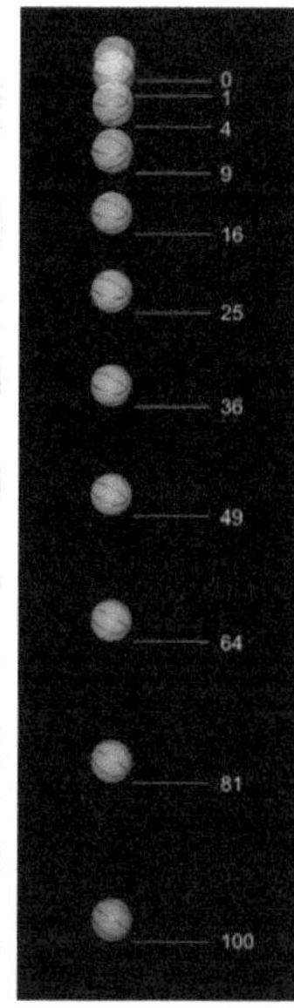

Ball falling freely under gravity. See text for description.

Under the same constant gravity assumptions, the potential energy, E_p, of a body at height h is given by $E_p = mgh$ (or $E_p = Wh$, with W meaning weight). This expression is valid only over small distances h from the surface of the Earth. Similarly, the expression for the maximum height reached by a vertically projected body with velocity v is useful for small heights and small initial velocities only.

Gravity and astronomy

The discovery and application of Newton's law of gravity accounts for the detailed information we have about the planets in our solar system, the mass of the Sun, the distance to stars, quasars and even the theory of dark matter. Although we have not traveled to all the planets nor to the Sun, we know their masses. These masses are obtained by applying the laws of gravity to the measured characteristics of the orbit. In space an object maintains its orbit because of the force of gravity acting upon it. Planets orbit stars, stars orbit Galactic Centers, galaxies orbit a center of mass in clusters, and clusters orbit in superclusters. The force of gravity is proportional to the mass of an object and inversely proportional to the square of the distance between the objects.

Gravitational radiation

In general relativity, gravitational radiation is generated in situations where the curvature of spacetime is oscillating, such as is the case with co-orbiting objects. The gravitational radiation emitted by the Solar System is far too small to measure. However, gravitational radiation has been indirectly observed as an energy loss over time in binary pulsar systems such as PSR B1913+16. It is believed that neutron star mergers and black hole formation may create detectable amounts of gravitational radiation. Gravitational radiation observatories such as LIGO have been created to study the problem. No confirmed detections have been made of this hypothetical radiation, but as the science behind LIGO is refined and as the instruments themselves are endowed with greater sensitivity over the next decade, this may change.

Anomalies and discrepancies

There are some observations that are not adequately accounted for, which may point to the need for better theories of gravity or perhaps be explained in other ways.

- **Extra fast stars**: Stars in galaxies follow a distribution of velocities where stars on the outskirts are moving faster than they should according to the observed distributions of normal matter. Galaxies within galaxy clusters show a similar pattern. Dark matter, which would interact gravitationally but not electromagnetically, would account for the discrepancy. Various modifications to Newtonian dynamics have also been proposed.

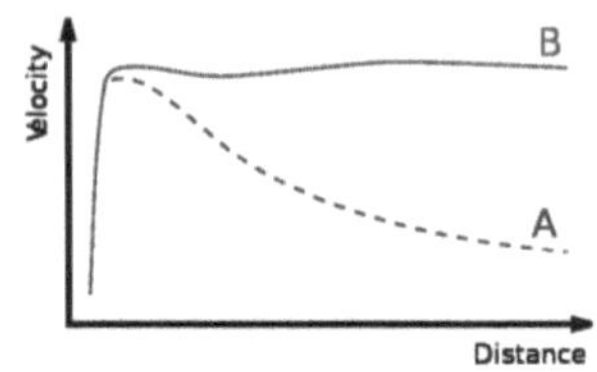

Rotation curve of a typical spiral galaxy: predicted (**A**) and observed (**B**). The discrepancy between the curves is attributed to dark matter.

- **Flyby anomaly**: Various spacecraft have experienced greater accelerations during slingshot maneuvers than expected.
- **Accelerating expansion**: The metric expansion of space seems to be speeding up. Dark energy has been proposed to explain this. A recent alternative explanation is that the geometry of space is not homogeneous (due to clusters of galaxies) and that when the data are reinterpreted to take this into account, the expansion is not speeding up after all,[21] however this conclusion is disputed.[22]
- **Anomalous increase of the astronomical unit**: Recent measurements indicate that planetary orbits are widening faster than if this were solely through the sun losing mass by radiating energy.
- **Extra energetic photons**: Photons travelling through galaxy clusters should gain energy and then lose it again on the way out. The accelerating expansion of the universe should stop the photons returning all the energy, but even taking this into account photons from the cosmic microwave background radiation gain twice as much energy as expected. This may indicate that gravity falls off *faster* than inverse-squared at certain distance scales.[23]
- **Dark flow**: Surveys of galaxy motions have detected a mystery dark flow towards an unseen mass. Such a large mass is too large to have accumulated since the Big Bang using current models and may indicate that gravity falls off *slower* than inverse-squared at certain distance scales.[23]
- **Extra massive hydrogen clouds**: The spectral lines of the Lyman-alpha forest suggest that hydrogen clouds are more clumped together at certain scales than expected and, like dark flow, may indicate that gravity falls off *slower* than inverse-squared at certain distance scales.[23]

Alternative theories

Historical alternative theories

- Aristotelian theory of gravity
- Le Sage's theory of gravitation (1784) also called **LeSage gravity**, proposed by Georges-Louis Le Sage, based on a fluid-based explanation where a light gas fills the entire universe.
- Ritz's theory of gravitation, *Ann. Chem. Phys.* 13, 145, (1908) pp. 267-271, Weber-Gauss electrodynamics applied to gravitation. Classical advancement of perihelia.
- Nordström's theory of gravitation (1912, 1913), an early competitor of general relativity.
- Whitehead's theory of gravitation (1922), another early competitor of general relativity.

Recent alternative theories

- Brans–Dicke theory of gravity (1961)
- Induced gravity (1967), a proposal by Andrei Sakharov according to which general relativity might arise from quantum field theories of matter
- In the modified Newtonian dynamics (MOND) (1981), Mordehai Milgrom proposes a modification of Newton's Second Law of motion for small accelerations
- The self-creation cosmology theory of gravity (1982) by G.A. Barber in which the Brans-Dicke theory is modified to allow mass creation
- Nonsymmetric gravitational theory (NGT) (1994) by John Moffat
- Tensor-vector-scalar gravity (TeVeS) (2004), a relativistic modification of MOND by Jacob Bekenstein
- Gravity as an entropic force, gravity arising as an emergent phenomenon from the thermodynamic concept of entropy.

See also

- Anti-gravity, the idea of neutralizing or repelling gravity
- Artificial gravity
- Einstein–Infeld–Hoffmann equations
- Escape velocity, the minimum velocity needed to escape from a gravity well
- g-force, a measure of acceleration
- Gravitational induction
- Gravitational binding energy
- Gravity assist
- Gravity Recovery and Climate Experiment
- Gravity Research Foundation
- Gauss' law for gravity
- Jovian-Plutonian gravitational effect
- Kepler's third law of planetary motion
- Lagrangian point
- Mixmaster dynamics
- Newton's laws of motion
- n-body problem
- Pioneer anomaly
- Scalar theories of gravitation
- Speed of gravity
- Standard gravitational parameter
- Standard gravity
- Weightlessness

Notes

- Proposition 75, Theorem 35: p. 956 - I.Bernard Cohen and Anne Whitman, translators: Isaac Newton, *The Principia*: Mathematical Principles of Natural Philosophy. Preceded by *A Guide to Newton's Principia*, by I. Bernard Cohen. University of California Press 1999 ISBN 0-520-08816-6 ISBN 0-520-08817-4
- Max Born (1924), *Einstein's Theory of Relativity* (The 1962 Dover edition, page 348 lists a table documenting the observed and calculated values for the precession of the perihelion of Mercury, Venus, and Earth.)

Footnotes

[1] Ball, Phil (06 2005). "Tall Tales". *Nature News*. doi:10.1038/news050613-10.

[2] Galileo (1638), *Two New Sciences*, First Day (http://oll.libertyfund.org/?option=com_staticxt&staticfile=show.php?title=753&chapter=109891&layout=html&Itemid=27) Salviati speaks: "If this were what Aristotle meant you would burden him with another error which would amount to a falsehood; because, since there is no such sheer height available on earth, it is clear that Aristotle could not have made the experiment; yet he wishes to give us the impression of his having performed it when he speaks of such an effect as one which we see."

[3] *Chandrasekhar, Subrahmanyan (2003). *Newton's Principia for the common reader*. Oxford: Oxford University Press. (pp.1–2). The quotation comes from a memorandum thought to have been written about 1714. As early as 1645 Ismaël Bullialdus had argued that any force exerted by the Sun on distant objects would have to follow an inverse-square law. However, he also dismissed the idea that any such force did exist. See, for example, Linton, Christopher M. (2004). *From Eudoxus to Einstein—A History of Mathematical Astronomy*. Cambridge: Cambridge University Press. p. 225. ISBN 978-0-521-82750-8.

[4] Dittus, H; C. Lämmerzahl (PDF). *Experimental Tests of the Equivalence Principle and Newton's Law in Space* (http://www.zarm.uni-bremen.de/2forschung/gravi/publications/papers/2005DittusLaemmerzahl.pdf). .

[5] Paul S Wesson (2006). *Five-dimensional Physics* (http://books.google.com/?id=dSv8ksxHR0oC&printsec=frontcover&dq=intitle:Five+intitle:Dimensional+intitle:Physics). World Scientific. p. 82. ISBN 9812566619. .

[6] Haugen, Mark P.; C. Lämmerzahl (2001). *Principles of Equivalence: Their Role in Gravitation Physics and Experiments that Test Them*. Springer. arXiv:gr-qc/0103067. ISBN 978-3-540-41236-6.

[7] "Gravity and Warped Spacetime" (http://www.black-holes.org/relativity6.html). black-holes.org. . Retrieved 2010-10-16.

[8] http://laser.phys.ualberta.ca/~egerton/genrel.htm

[9] Pauli, Wolfgang Ernst (1958). "Part IV. General Theory of Relativity". *Theory of Relativity*. Courier Dover Publications. ISBN 9780486641522.

[10] Dyson, F.W.; Eddington, A.S.; Davidson, C.R. (1920). "A Determination of the Deflection of Light by the Sun's Gravitational Field, from Observations Made at the Total Eclipse of May 29, 1919". *Phil. Trans. Roy. Soc. A* **220** (571–581): 291–333. Bibcode 1920RSPTA.220..291D. doi:10.1098/rsta.1920.0009.. Quote, p. 332: "Thus the results of the expeditions to Sobral and Principe can leave little doubt that a deflection of light takes place in the neighbourhood of the sun and that it is of the amount demanded by Einstein's generalised theory of relativity, as attributable to the sun's gravitational field."

[11] Weinberg, Steven (1972). *Gravitation and cosmology*. John Wiley & Sons.. Quote, p. 192: "About a dozen stars in all were studied, and yielded values 1.98 ± 0.11" and 1.61 ± 0.31", in substantial agreement with Einstein's prediction $\theta_{\odot} = 1.75"$."

[12] Earman, John; Glymour, Clark (1980). "Relativity and Eclipses: The British eclipse expeditions of 1919 and their predecessors". *Historical Studies in the Physical Sciences* **11**: 49–85.

[13] Weinberg, Steven (1972). *Gravitation and cosmology*. John Wiley & Sons. p. 194.

[14] See W.Pauli, 1958, pp.219–220

[15] NASA's Gravity Probe B Confirms Two Einstein Space-Time Theories (http://www.nasa.gov/home/hqnews/2011/may/HQ_11-134_Gravity_Probe_B.html)

[16] Galaxy Clusters Validate Einstein's Theory (http://news.sciencemag.org/sciencenow/2011/09/galaxy-clusters-validate-einstei.html)

[17] Randall, Lisa (2005). *Warped Passages: Unraveling the Universe's Hidden Dimensions*. Ecco. ISBN 0-06-053108-8.

[18] Feynman, R. P.; Morinigo, F. B., Wagner, W. G., & Hatfield, B. (1995). *Feynman lectures on gravitation*. Addison-Wesley. ISBN 0201627345.

[19] Zee, A. (2003). *Quantum Field Theory in a Nutshell*. Princeton University Press. ISBN 0-691-01019-6.

[20] Greene, Brian (2000). *The elegant universe: superstrings, hidden dimensions, and the quest for the ultimate theory*. New York: Vintage Books. ISBN 0375708111.

[21] Dark energy may just be a cosmic illusion (http://space.newscientist.com/channel/astronomy/cosmology/mg19726461.600-dark-energy-may-just-be-a-cosmic-illusion.html), *New Scientist*, issue 2646, 7th March 2008.

[22] Swiss-cheese model of the cosmos is full of holes (http://space.newscientist.com/article/mg20026783.800-swisscheese-model-of-the-cosmos-is-full-of-holes.html), *New Scientist*, issue 2678, 18th October 2008.

[23] "Where Matter Fears to Tread", New Scientist issue 2669, 14 March 2009

References

- Halliday, David; Robert Resnick; Kenneth S. Krane (2001). *Physics v. 1*. New York: John Wiley & Sons. ISBN 0-471-32057-9.
- Serway, Raymond A.; Jewett, John W. (2004). *Physics for Scientists and Engineers* (6th ed.). Brooks/Cole. ISBN 0-534-40842-7.
- Tipler, Paul (2004). *Physics for Scientists and Engineers: Mechanics, Oscillations and Waves, Thermodynamics* (5th ed.). W. H. Freeman. ISBN 0-7167-0809-4.

Further reading

- Thorne, Kip S.; Misner, Charles W.; Wheeler, John Archibald (1973). *Gravitation*. W.H. Freeman. ISBN 0-7167-0344-0.

rue:Гравітація

Ground-penetrating radar

Ground-penetrating radar (**GPR**) is a geophysical method that uses radar pulses to image the subsurface. This nondestructive method uses electromagnetic radiation in the microwave band (UHF/VHF frequencies) of the radio spectrum, and detects the reflected signals from subsurface structures. GPR can be used in a variety of media, including rock, soil, ice, fresh water, pavements and structures. It can detect objects, changes in material, and voids and cracks.[1]

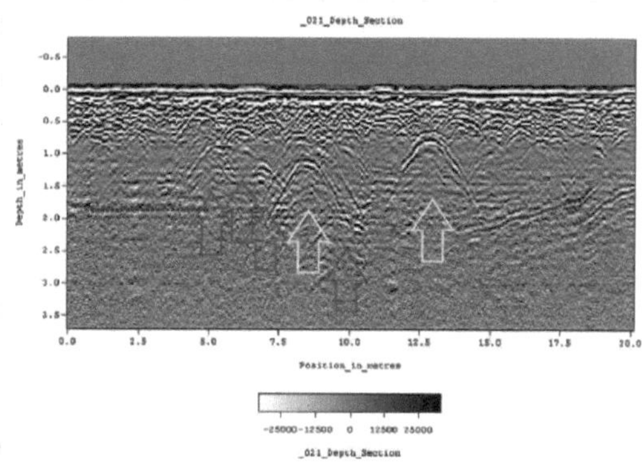

A ground-penetrating radargram collected on an historic cemetery in Alabama, USA. Hyperbolic reflections indicate the presence of reflectors buried beneath the surface, possibly associated with human burials.

GPR uses high-frequency (usually polarized) radio waves and transmits into the ground. When the wave hits a buried object or a boundary with different dielectric constants, the receiving antenna records variations in the reflected return signal. The principles involved are similar to reflection seismology, except that electromagnetic energy is used instead of acoustic energy, and reflections appear at boundaries with different dielectric constants instead of acoustic impedances.

The depth range of GPR is limited by the electrical conductivity of the ground, the transmitted center frequency and the radiated power. As conductivity increases, the penetration depth decreases. This is because the electromagnetic energy is more quickly dissipated into heat, causing a loss in signal strength at depth. Higher frequencies do not penetrate as far as lower frequencies, but give better resolution. Optimal depth penetration is achieved in ice where the depth of penetration can achieve several hundred meters. Good penetration is also achieved in dry sandy soils or massive dry materials such as granite, limestone, and concrete where the depth of penetration could be up to 15 m. In moist and/or clay-laden soils and soils with high electrical conductivity, penetration is sometimes only a few centimetres.

Ground-penetrating radar antennas are generally in contact with the ground for the strongest signal strength; however, GPR air launched antennas can be used above the ground.

Cross borehole GPR has developed within the field of hydrogeophysics to be a valuable means of assessing the presence and amount of soil water.

Applications

GPR has many applications in a number of fields. In the Earth sciences it is used to study bedrock, soils, groundwater, and ice. Engineering applications include nondestructive testing (NDT) of structures and pavements, locating buried structures and utility lines, and studying soils and bedrock. In environmental remediation, GPR is used to define landfills, contaminant plumes, and other remediation sites, while in archaeology it is used for mapping archaeological features and cemeteries. GPR is used in law enforcement for locating clandestine graves and buried evidence. Military uses include detection of mines, unexploded ordnance, and tunnels.

Ground penetrating radar in use. Near Stillwater, OK, 2010

Before 1987 the Frankley Reservoir in Birmingham, England UK was leaking 540 litres of drinking water per second. In that year GPR was used successfully to isolate the leaks.[2]

Borehole radars utilizing GPR are used to map the structures from a borehole in underground mining applications. Modern directional borehole radar systems are able to produce three-dimensional images from measurements in a single borehole.

One of the other main applications for ground penetration radars to locate underground utilities, since GPR is able to generate 3D underground images of pipes, power, sewage and water mains.

Ground penetrating radar survey of an archaeological site in Jordan.

Three-dimensional imaging

Individual lines of GPR data represent a sectional (profile) view of the subsurface. Multiple lines of data systematically collected over an area may be used to construct three-dimensional or tomographic images. Data may be presented as three-dimensional blocks, or as horizontal or vertical slices. Horizontal slices (known as "depth slices" or "time slices") are essentially planview maps isolating specific depths. Time-slicing has become standard practice in archaeological applications, because horizontal patterning is often the most important indicator of cultural activities.

Limitations

The most significant performance limitation of GPR is in high-conductivity materials such as clay soils and soils that are salt contaminated. Performance is also limited by signal scattering in heterogeneous conditions (e.g. rocky soils).

Other disadvantages of currently available GPR systems include:

- Interpretation of radargrams is generally non-intuitive to the novice.
- Considerable expertise is necessary to effectively design, conduct, and interpret GPR surveys.
- Relatively high energy consumption can be problematic for extensive field surveys.

Recent advances in GPR hardware and software have done much to ameliorate these disadvantages, and further improvement can be expected with ongoing development.

Power regulation

In 2005, the European Telecommunications Standards Institute introduced legislation to regulate GPR equipment and GPR operators to control excess emissions of electromagnetic radiation [3] . The European GPR association (EuroGPR) was formed as a trade association to represent and protect the legitimate use of GPR in Europe.

Similar technologies

Ground penetrating radar uses a variety of technologies to generate the radar signal, these are impulse, stepped frequency, FMCW and noise. Systems on the market in 2009 also use DSP to process the data, while survey work is being carried out rather than off line.

There is also a special kind of GPR that uses unmodulated continuous-wave signals. This is holographic subsurface radar that differs from other GPR types in that it records plan-view subsurface holograms. Depth penetration of this kind of radar is rather small (up to 20-30 cm), but lateral resolution is enough to discriminate different type of landmines in the soil or cavities, defects, bugging devices, or other hidden objects in walls, floors and structural elements.

GPR is used on vehicles for close-in high speed road survey and landmine detection as well as in stand-off mode.

Pipe Penetrating Radar (PPR) is an application of GPR technologies applied in-pipe where the signals are directed through pipe and conduit walls to detect pipe wall thickness and voids behind the pipe walls.

Wall-penetrating radar can read through walls and even act as a motion sensor for police.

GPR is used for hand held landmine detection to reduce the false alarms experienced by the standard metal detector and systems are available off the shelf (Vallon and L3Com Cyterra)

The "Mineseeker Project" seeks to design a system to determine whether landmines are present in areas using ultra wideband synthetic aperture radar units mounted on blimps.

References

[1] Daniels DJ (ed.) (2004). *Ground Penetrating Radar* (2nd ed.). Knoval (Institution of Engineering and Technology). pp. 1–4. ISBN 978-0-86341-360-5.

[2] Penguin Dictionary of Civil Engineering p347 (*Radar*)

[3] ETSI EG 202 730 V1.1.1 (2009-09), "Electromagnetic compatibility and Radio spectrum Matters (ERM); Code of Practice in respect of the control, use and application of Ground Probing Radar (GPR) and Wall Probing Radar (WPR) systems and equipment

Borchert, Olaf: *Receiver Design for a Directional Borehole Radar System* Dissertation, University of Wuppertal, 2008, (http://nbn-resolving.de/urn/resolver.pl?urn=urn:nbn:de:hbz:468-20080382)

External links

- EUROGPR – The European GPR regulatory body (http://www.eurogpr.org)
- GprMax – GPR numerical simulator based on the FDTD method (http://www.gprmax.org)
- Short movie showing acquisition, processing and accuracy of GPR readings (http://www.youtube.com/watch?v=RHVyERIB-to)

Article Sources and Contributors

Frankley Reservoir *Source*: http://en.wikipedia.org/w/index.php?title=Frankley_Reservoir *Contributors*: Aatomic1, D6, Docu, Glen, Grutness, Hydrogen Iodide, Pigsonthewing, SP-KP, Trident13, Valiantis, 4 anonymous edits

Reservoir *Source*: http://en.wikipedia.org/w/index.php?title=Reservoir *Contributors*: Aatox, Acalamari, Alansohn, Allstarecho, AmericanLeMans, Amniarix, Andonic, Arjayay, Arthena, Astirmays, Awickert, B.d.mills, BL Lacertae, Babakathy, Bassbonerocks, Billinrio, Blehfu, Bobo192, Brougham96, Camw, Can't sleep, clown will eat me, CanadianLinuxUser, Camildo, Cassowary, Cctoide, Ceranthor, Chl, Chris j wood, ChrisCork, Ciphers, CommonsDelinker, Courcelles, Csmart287, CustardJack, Cyfal, DFS454, DVD R W, Davandron, Deror avi, DidiWeidmann, Dj Capricorn, Dnj710, Docu, Donarreiskoffer, Drunken Pirate, ENeville, ESkog, Em3rald, Emerson7, Eranjenes2, Erik9, Ewen, Ezhiki, FayssalF, Fieldday-sunday, Fifth Fish Finger, Floul1, Frehley, Fribbler, Frokor, Funandtrvl, Fæ, GLaDOS, Gail, GavinSharp, Gcm, GeoWriter, Geoff Bonanno, Geonarva, Gersenes, Ghirlandajo, Gilliam, HarlandQPitt, Hertz1888, HexaChord, Hike395, Hires an editor, Hmains, Ian n-s, Infrogmation, Ingolfson, J.Gifford, JBellis, JCDenton2052, JForget, Jacksieber, JamesAM, Jeffrey Mall, JetLover, Jim.henderson, Jonpollnow, Jose Icaza, Jpbowen, KVDP, Kaiba, Karlos87, Kbthompson, Kimse, Kjetil r, Kjkolb, Kostisl, Krj373, Kschwerdt514, Levineps, Lobob5, Logan, Lysy, Mailseth, Makeemlighter, Malathos, Marcia Wright, Martial75, Mattgirling, Mattisse, MaxSem, MeekMark, Metropolitan90, MichaelMaggs, Minghong, Mircea cs, Moanzhu, Moreau1, Muijz, Muéro, Narson, Newell Post, Nigholith, Nihiltres, Nixsun01, Nk, NortyNort, Nshimbi, Ohdear15, Oismiffy, OlEnglish, Old Moonraker, Orphan Wiki, Osarius, Pengo, Pete ess, Peter gk, Peterlewis, Philip Trueman, Pigsonthewing, Planetjanet, Pointillist, PopUpPirate, Prabinepali, Prashanthns, Qfl247, RJFJR, Radiojon, Rajasekhar1961, Ravedave, Redmegtheavenger, Rich Farmbrough, Robina Fox, Roxi2, Rruss, S4n1HS22WMX695In, Samnz05, SchuminWeb, Sean the Spook, Severnapark, Sfahey, Shannon1, Sole Soul, Squids and Chips, Svetovid, Swanml, Tabletop, Tedickey, TenOfAllTrades, TheGerm, Thingg, Thw1309, Tide rolls, Tills, Tom harrison, Totse-vandal, Tranquil demon, Trevyn, Trilby*foxglove, Tsuite, Túrelio, Uncle Dick, Unschool, VPliousnine, Vdegroot, Vegaswikian, Velella, Versus22, Vhorvat, VinceBowdren, Wafulz, Walrusfunk, Wavelength, Whosasking, Wikid77, Wikijmms, William Avery, YixilTesiphon, 246 anonymous edits

Drinking water *Source*: http://en.wikipedia.org/w/index.php?title=Drinking_water *Contributors*: 16@r, A8UDI, AS, Abtinb, Adamj0123, Aerotwelve, Afed, Agateller, Agoldstand, Aguyonearth, Ahuskay, Alansohn, Aled D, Alex vnsc, AlexWaelde, AlexiusHoratius, Altair786, Amokbel, Anclation, AndrewHowse, Andy Smith, AngelLeliel, Anlace, AnneDELS, Ariel., Arthur Rubin, AstroHurricane001, Auntof6, Avenged Eightfold, Avono, Babando, BanyanTree, Baseballbaker23, Becritical, Beetstra, Benjihiro, Beta34, Biscuittin, Bjelkeman, Blueiris, BoP, BobbyPeru, Bobo192, Bockbockchicken, Brian the Editor, Bueller 007, Bugo30, BullRangifer, CDM2, CanadianLinuxUser, Capricorn42, Carrionluggage, Cburnett, Ccacsmss, Cdcdoc, Chamdarae, Cherylrenee, Chochopk, Chris the speller, ClosedEyesSeeing, Closedmouth, Cocytus, Colonel Warden, Covalent, Cydperez, DARTH SIDIOUS 2, DMCer, Darkwind, Dawn Bard, DeadEyeArrow, Decltype, Deli nk, Demiurge, Deror avi, Diarrhoea0176556, Dieselbub, Dj Capricorn, Docu, DonVander, Doug Coldwell, Dr.frog, Dr.jph, DuncanHill, Dust Filter, E Wing, EJF (huggle), Ebear422, Edgar181, Ejhoekstra, El C, El Rutabaga, Elliskev, Energelproducts, Epbr123, Erdinal, Erianna, Eric-Wester, Eric567, Ethan, Eug.galeotti, Everyking, Exxolon, FOK SD OA, Faradayplank, FayssalF, Fieldday-sunday, Flic182, Fourdee, Freiberg, Gabbe, Gabriel Kielland, Gaius Cornelius, GaiusTimiusAwesomus, Gak, Gamaliel, Garion96, Generalcp702, Ggjacobsen, Giftlite, Ginkgo100, Glane23, Gold79, Graham87, Greenknight21, Grundle2600, Gwernol, Gyrofrog, Hankwang, Hdt83, HealthywaterCDC, Heero Kirashami, Heron, HexaChord, HisSpaceResearch, Hmains, Hu12, Ian.bjorn, ImperfectlyInformed, Ionut Cojocaru, Iridescent, Ironman.james, J.delanoy, JBellis, JCDenton2052, JDCMAN, Jack12345673, Jackollie, Jacob Lundberg, JamesAM, Jamestttgrays, Jan1nad, Japanese Searobin, Jeff G., Jeffalbert, Jehochman, Jemiller226, Jersyko, Jimjamjak, Jmundo, Joan-of-arc, John Broughton, John Reiher, John254, Johntindale, Johnuniq, Jonathanmcguinness, Joseph Bullrich, Jreconomy, Jwh, KVDP, Kajerm, Keilana, Kelvinc, Khoikhoi, Khooly59, KieferSkunk, Konaallan, Ksyrie, Kurt Shaped Box, Kuru, KyjL, La Pianista, Larry Yuma, Les boys, Leslie Mateus, Liftarn, Longhair, Loren.wilton, Loverbunny4u, Ltholen, Luna Santin, Lyoudale, MER-C, Mabu2, MacStep, Maineiac2569, Mandarax, Martin Chaplin, MasterOfHisOwnDomain, Materialscientist, Maxí, Mbeychok, Megapixie, Mercuryrisen, Michael Devore, Michael Marcus, Michael R Bax, Michaelas10, Michaelfoti, Mikael Häggström, Mild Bill Hiccup, MindlessXD, Mitsukai, Mkweise, Moreau1, Mr Bungle, MrOllie, Mrs Trellis, Mschiffler, Murzilka, Mwanner, Mymagritte, Mzuo, N0d3, NOLA504ever, NPIC, Naturalenvironment2011, NellieBly, Neverquick, NewEnglandYankee, Nil Einne, Niluop, Nivix, Noq, Nukeless, OKtosiTe, ONUnicorn, Officiallyover, Ohnoitsjamie, Padolsey, Patrick987, Pbeesley1989, Persian Poet Gal, Peter Horn, Peterson.Amy.E, Pethr, Pigsonthewing, Pinethicket, Positron, Pseudomonas, R'n'B, Radagast83, Ratiocinate, Raygunnerson, Realm of Shadows, Reaper Eternal, RedSpruce, Redthoreau, Renee961, RevolverOcelotX, Reywas92, Rick DeLong, Rivyolie, Rjwilmsi, Rodriguez2321, Rrenner, Rror, Rtc, RyanGerbil10, S8946273h, SLIVER2964, Safewater, Sap123, Sapphirelove, Sarahg27, Saturday, Sekolov, Senator Palpatine, SheepNotGoats, Shinkolobwe, Shoeshinecs, Siddhant, Silencedmajority, Silvia879, SiobhanHansa, Skipsievert, Skunkboy74, Slakr, Slightsmile, Smallman12q, Smarterh2o, Smokefoot, Spiffy sperry, Spitfireatme, Stadtpark, Stone, Strongsauce, Super-Magician, Sweetpaseo, Switpi84, Symane, SyntaxError55, Syrthiss, T He He, Tabletop, Taishatjye, Tamás Kádár, Tangotango, Tarret, Tekleni, Temanning, The Thing That Should Not Be, ThePoopSmith, Thu, Thumperward, Tide rolls, TidusBlade, Tkn20, Toddst1, Toiyabe, Tom Pippens, Tom Sponheim, Tomeasy, Trainra, TurboChan, U235, USchick, V.narsikar, Valerius Tygart, Vegaswikian, Veinor, Velella, Visarequirements, Vmenkov, Vsmith, WCFrancis, WDickens, Wahrmund, Warpflyght, Warrior4truth, WatermelonPotion, Watertruth, Wavelength, Weetoddid, WhatamIdoing, WikHead, Wikieditor06, WikipedianMarlith, Will Beback, William Avery, Wowomg.ca, Wperdue, Wsiegmund, X!, Xero.Ye, Xiahou, Xyzzyplugh, Yworo, Zidonuke, Zodon, Zyresic, Zzuuzz, 686 anonymous edits

Birmingham *Source*: http://en.wikipedia.org/w/index.php?title=Birmingham *Contributors*: *drew, 0Andrew0, 1exec1, 2cools, 3WildMon, 54gsze4ghz5, 80.255, A bit iffy, A. Parrot, AEMoreira042281, AJMW, AJR, Aarandir, Aatho09, Ab9, Absolutadam802, Acalamari, Achangeisasgoodasa, Acs4b, Adam tomes, Admittingfear, Aecis, Aembleton, Aeusoes1, AgarwalSumeet, Ahoerstemeier, Air luigi, Ajh16, Akuru, AlabamaGuy2007, Alai, Alanbuchanan, Alansohn, Alarics, Aleenf1, Alkivar, AllanHainey, Allstarecho, Alun009, Alzarian16, Amakuru, Amatire, Amazonien, Ancient Apparition, Andre Engels, Andrew Norman, AndrewHowse, AndrewRT, Andrewd-t, Andy G, AndyPryke, Andyjsmith, Angela, Angmering, Angr, Angusmclellan, Aniten21, Antandrus, Aquilina, Archivist, Arif Guder, Arjun01, Arnfieldroad, Arpingstone, Arru111, Art LaPella, Arthena, Artlondon, Arwel Parry, Asdfasdf1231234, Asfasgad, Asia Minor, Asidemes, Astrotrain, Atleeek, Atrevers, AtticusX, Attilios, Aude, Auximines, Avala, Avaquava1, Avenue, Avnjay, AxG, AyanP, Aymatth2, Aynuk N. Ayli, BD2412, Babyoil, Backslash Forwardslash, Bad.heron, Badapro, Bagofants, Balalayker, Balrog-kun, Barbara Shack, Barnabypage, Bblfan, Bchapas, Bdog9121, Bearcat, Beek man, Beetstra, Belovedfreak, BenShade, Bento00, Benzedrine Clown, Beomora, Berek, Betacommand, Betterusername, Beve, BexTech, Bhadani, Bhamdoll, Biker Biker, Binbaby, Bissinger, Bkell, Bkonrad, Black Falcon, Black Kite, Blade4life1727, Blindbasher, Blood Red Sandman, Bluelion, Bluestarone, Bobblewik, Bobfish 341, Bobfos, Bobo192, Bobspeltbackwards, Bonadea, Brandon1978, Brianlucas, Brum=scum, BrumBoy1234, Brumbeat, Brumburger, Brumwalsal, Brusselsshrek, Bryan.burgers, Bs0u10e01, Burtonupontrent, Buxbaum666, Bwithh, C0pernicus, C4mp8311, CBDunkerson, Cabinscooter, Cactus.man, Camembert, Can't sleep, clown will eat me, CanisRufus, Canterbury Tail, Captain-tucker, Cassowary, CatherineMunro, Caulde, Cavrdg, Cecropia, Cfailde, Cgadavies, Chanheigeorge, Charleswiles, CharlotteWebb, Ched Davis, Chivista, Choalbaton, Chocolatechipcookie91, Chowells, Chris Roy, ChrisJB, ChrisTheDude, Chrisbrum, Chrislk02, Christian75, Christoff2k7, Cimon Avaro, Ck lostsword, Ckatz, Clare.s.hudson, Clay70, Cleverlew, Closedmouth, Cls14, Cockmaster69, Colin Marquardt, Colinpalmer, College35, Cometstyles, CommonsDelinker, Conscious, Cordless Larry, Cormaggio, Corvus cornix, Counsell, Courcelles, Courtinjustice, Crablogger, Craig Stuntz, Creativemidlands, Crunk, Cs-wolves, Cst17, Curran1980, CustardJack, Cutler, D6, DARTH SIDIOUS 2, DJ Clayworth, DON'T HAVE A BLOW OUT!!!, DScarface1991, DWaterson, DaDoc540, Daduzi, Daledo123, Damian805, Dancingwombatsrule, Daniel, Daniel625, DarkFalls, Darthgamer6621, DarwinPeacock, Dave.Dunford, Davelong, David Gerard, David john bonnett, Davyboyuk, Dawidl, Dcoetzee, DeadEyeArrow, Debresser, Deflective, Deltabeignet, Demograph, DerBorg, DerHexer, Dergraaf, Deus Ex, Discomike, Discospinster, DivineIntervention, Djdannyp, Djidash, Dm1987, Dmn, DocendoDiscimus, Docu, Doczilla, Dogaroon, Doric Loon, Doulos Christos, Dpaajones, Dpmuk, Dr bab, Dr. Blofeld, Dravecky, Dreadarthur, Dreaded Walrus, Dreadstar, Dryazan, Duja, Duncharris, Dyaimz, Dysprosia, EarlyBird, Ecotrend, Ed g2s, EdC, Edward, Efghij, Egghead06, Eh85, El C, Emesjee, Enable Admission, Enchanter, Enochlau, Enzedbrit, Epa101, Epbr123, Epolk, Erebus555, Eric119, Eric82oslo, Erik Baas, Escape Orbit, Esperant, European Girl, Eurosong, Everyking, Explicit, FCM2010, Fallschirmjäger, Farhan Murtaza, Fastestdogever, Fastfission, Felix Folio Secundus, Fingerpuppet, Finlay McWalter, Fish and karate, Flash94, Flatterworld, Fluffy McNutter, Folantin, Fontboy, Fontgirl, Foobar, Fourohfour, Francs2000, Frazzydee, Fredd the Redd, Freddiecaveman, Freddy lunch, Fredrik, FreplySpang, Frisco-98, Fruity Gonzalez, Funandtrvl, Funnyhat, Fuzheado, Fyremad!, Fæ, G-Man, GRAB SOME JEYES FLUID!, GRAHAMUK, GRB1972, GTBacchus, Gabbe, Gailtb, Gaius Octavius Princeps, Galati, Galoubet, Gammondog, GarryPaterson1993, Garyjillgolf, Gbwikiniki, Gdr, Geezer hd1, Gem, Geneb1955, General Eisenhower, Geopersona, Georgequizbowl08, Getafix, Ghostwords, GilesSutcliffe, Gillean666, Givesort, Glass Sword, Globtik1, GorillaWarfare, Goudzovski, GraemeL, Grafen, Graham87, Grarhl123, GreeceGroove, Green Wolf Cub, Greenshed, GregAsche, GregRobson, Grover cleveland, Grunners, Grunt, Gsl, Gunnernett, Guntramshatterhand, Gurch, Guy Peters, Gwernol, Hadseys, Halo, Ham, Hamedog, Hamilton365, Hanad9395, Hassocks5489, Hayden120, Hayley and them peas eh, Hcheney, HelloGoofballs, Heron, HexaChord, Highvale, Historicist, Hoary, Hobsholelane06, Holtville, Hotlorp, Howesey, Hu, Hughpugh, Hurricane1978, Husond, I THINK YA NEED A CORK UP THERE TO STOP THE FLOW!, Iain99, Iainscott, Ian G Collins, Ian3055, Ianb, Icairns, Icebergslim, Icseaturtles, Ifibabe, Ify00, Ilikeeatingwaffles, Imperium Europeum, InfectedWithRage, Infrogmation, Intelitech, Ipeqi, Irate, Irfghanistani, It's mi tablets I swear!, J.delanoy, JBellis, JFRogers, JForget, JHarrisTodd, Jaberwocky6669, Jackfork, Jambo560, James Dowden, James Emtage, James26, Jamesasmall, Jamesinderbyshire, Jan1nad, Jatkins, Jauhienij, Jay-Sebastos, Jayglamar, Jbmurray, Jdforrester, Jecar, Jeff3000, Jehuty102, Jeni, Jeremy Rowett Johns, JeremyA, Jeronimo, Jhamez84, JimmyGuano, Jimtaip, Jkelly, Jlh204, JoanneB, Joe.dickens, Joe9y, JoeSmack, Joewithajay, John, Johndarrington, Jon Cates, Jonathan de Boyne Pollard, Jonathunder, Jonel, Jonik, Jonny1969wood, Joolz, Jose77, Joseph1066uk, Joshii, Jpb1301, Jpers36, Jredmond, Jrobin08, Jsc83, Juglugs, Juliancolton, Junegirl, JustAGal, Justinc, Jwrosenzweig, JzG, Jza84, Ka Faraq Gatri, Kaihsu, Karl1587, Karl2620, Karlos87, Katalaveno, Kate, KathRUNE, Kavanagh21, Kbdank71, Kedi the tramp, Keilana, Keith D, Keith Edkins, Kelly Martin, Kenaldinho10, Kenwarren, Kevinwiki999, Keyroon13, Keysmall, Khanda911, Khendon, Kieron a m, Killagb, Kim Bruning, King konger, King of Mercia, Kingpin13, Kirrages, Kjlewis, Kmrayner, KnightLago, KnowledgeOfSelf, Knucmo2, Koavf, Korg, Kotek007, Kozuch, Kraftlos, Kroum, Kusma, Kwamikagami, LSky, Lankyboy, LarRan, Latiosman1, Laurel Lodged, LawBham, Le Fou, Legge1996, Leonig Mig, Lerdsuwa, Lfh, LiDaobing, Libertyville, Lightmouse, LilHelpa, Lilac Soul, LindsayH, LionOfJudah80, LittleOldMe, Locke Cole, Lofty, Logan, Logical Cowboy, Lohengrin1991, Loltest123, Lord Emsworth, Lotje, Loudrocksurfer, Lovelyspottedcow, Lozleader, Lozzaboy, Lst27, Luna Santin, Lupo, M 1132, MER-C, MJCdetroit, MONGO, MRSC, Magnalord, Magnus Manske, Majabl, Majorly, Malcolm, Malcolm Farmer, Man vyi, Manchester2k6, Mancunianboy, MapsMan, Marcus22, Marek69, Mark J, MarkSmith19222, MarkThomas, Marknew, Marky-Son, Marnanel, Marriex, Martarius, Martin451, MastCell, Materialscientist, Mathpianist93, Mattbr, Mattgirling, Mattley, Mav, Max olwen, Maxamegalon2000, Maxim, Maximus Rex, Mbc362, McTabish, Meaghan, Mel Etitis, Melburnian, Mellin avfc, Mentifisto, Metabaronic, Mgp28, Mholland, Mic, Michael Hardy, MightyWarrior, Mikale, MikeF9, MikeLynch, Mikemusson, Mild Bill Hiccup, Minesweeper, Mintguy, Missing Ace, Modster, Monglettuce, Monkeysee75, Moocha, Moondyne, Mortense, Morwen, Mossley10, Moswento, Mouse Nightshirt, Mr Brummie, Mr Stephen, MrAshJam, MrHaiku, Mrceleb2007, Mrmuk, Munci, MusicGeek101, Mybox, Mysitced, Mysticed, Mzilikazi1939, NJW494, Nadavspi, Naddy, Nagytibi, Napier jack, NawlinWiki, Nemilar, Neutral-being, Neutrality, Nev1, Nevilley, NewMember10go10by10me, Newguernsey, Newsheep, Nick Boulevard, Nickthenoo, Nikai, Nikki311, Ninestiles, Niteowlneils, No Guru, Noblesense, Noisy,

NoisyJinx, Nopetro, Npd2983, Numb3r3, Nyar94, Nydigoveth, Nyttend, Ocean Bacon, Odie5533, Ohconfucius, Ohnoitsjamie, Old Moonraker, Oliver Chettle, Omarraii, Omicronpersei8, One wit green hat, Oosoom, Opalstar, Opelio, Open2universe, OpenToppedBus, Optimist on the run, Orange Suede Sofa, Orangejon, Oreo Priest, Orphanannieathurslad, Otduff, Overminor, Owain, Oxydo, P.K., PGWG, PJ101300, Paldred, PamD, Parable1991, Paranomia, Patcat88, Pavel Vozenilek, Pcb21, Pcpcpc, Peaceworld111, Pebbles1979, Peregrine981, Peripitus, Petedk42, Peter, Peter v, PeterHuntington, Pharaoh of the Wizards, Phi beta, Philip Cross, Philip Trueman, Philthecow, Phlebas, Phoenixrod, Picapica, Picaroon, Pigsonthewing, Pip92, Plastikspork, Poindexter Propellerhead, Politepunk, Politics rule, Popsracer, Portnadler, Postlebury, Prashanthns, Princess Tiswas, ProCivitate, Prolifik, Proxima Centauri, PseudoOne, Psidogretro, Punkromance, PurpleHeather, Pyrotec, Qhgslcmcdyudgeis, Quantpole, Qxz, R'n'B, R. fiend, RFBailey, RHaworth, RRJ, Radagast, Rajput m16, RandomP, RandomXYZb, Raul654, Ravenhurst, Raygirvan, Rayoflight278, Rbrwr, Realist2, RedWolf, Redrug1, Redvers, Reedy, Rees11, Reginmund, Renesis, RetiredUser2, Rex the first, RexNL, Reywas92, Rholton, Rhyddfrydol, Rich Farmbrough, RichF, Richard Harvey, Richard barnes, RichardTimewarp, Richardbates2002, Rick Block, RickK, Ricky81682, Right blowout, Ringbark, Rjd.1892, Rjwilmsi, Rlfb, Robert of Ramsor, RobertG, Rockerbaby, Rockybiggs, Rockylaneman, RodCrosby, Romney yw, Romper, Roots2k, Roundhouse0, Roxi2, Rror, Rsrikanth05, Rtdixon86, RyanGerbil10, SP-KP, Saga City, Salaka, Saleskeeps, Sam Francis, Sam Hocevar, Samrica, Sand Patrol, SarekOfVulcan, ScatFace, Sceptre, SchfiftyThree, Schumi555, Sciurinæ, Scottrb, Sdm2211, Sdrawkcab, SeanMack, Selket, Serius123tjb, Setwisohi, Shantavira, Shizzy1, Showbizkid, Shropman, Shyguy1991, Sietse Snel, SilentVicinity, Silivrenion, SimonD, Simonk133, Simpaticante, Simply south, Sir Stanley, Sjorford, Sjstafford32, Skier Dude, Skodajon, Sladen, Sleepingsatelite, Slyguy135, Smalljim, Smallweed, Smartadvice, Smithy2471, Smokizzy, Snakephantom, Snowded, Snowmanradio, Solipsist, Sophus Bie, Southamite, Spartaz, Spearhead, Speedboy Salesman, Spellmaster, Spiby, Splash, SpookyMulder, Sprigot, Srm07, Sswonk, Star-one, Station1, Steel, SteinbDJ, Steinsky, Stephenb, Stetay, Stevo1000, Stevvvv4444, Stormie, StormySan, Strike2027, Strollers, Stroppolo, Suave12, Sugarshoesfootwear, SupXl, Super Mario, SuspendedSky, Swarm, Sweetcorn n all!, TFoxton, THELEICESTERFOX, Tagishsimon, Taoiseach, Tarquin, Tassedethe, Tawker, Tbone762, Teambanana, Technobliterator, TehGrauniad, Template namespace initialisation script, Texture, ThaddeusB, TharkunColl, Thatotherdude, The Anome, The Nut, The Random Editor, The sensei of truth, The wub, TheBoogler, TheDevilWearsNike, Thebogusman, Thehelpfulone, TheoClarke, Thewebeditor, Thingg, Think outside the box, ThinkerThoughts, Thommy-86, Thryduulf, Tiller54, Tim!, Timc, Timcoles29, Timillustrator, Timwi, Tnen, Tobby72, TodorBozhinov, Toi, Tom-, TomB123, TomParkerAVFC, Tomasfoster, Tony Fox, Tony Sidaway, Tony1, TonyW, Topbanana, Towerblocktom, Tpbradbury, Trackster, Travelbird, Trawicks, Tresiden, Trilobite, Triona, Tripod86, Truco X, Ttownfeen, Tud123, Turkstud, Tuxraider reloaded, Typicalmale, UK 007, UKbandit, Ugur Basak, Ukbrumboy, Ukweatherdataman, Ultrabeater, Uncle Dick, Uncle G, Unknown Unknowns, Unreal7, Uriber, User123tjb, Usgnus, Uxejn, Valenciano, Valiantis, Vapour, Vbrum, Velvettfogg, Ventusa, Vicki Rosenzweig, Viewfinder, Villafancd, Villafanuk, Violncello, Voretus, Vox Humana 8', Völva, WOSlinker, Walshy86, Wanipk, WarioLoaf, Warofdreams, Wasted Time R, Watch37264, WebSufer, Welsh, Welshleprechaun, Wereon, WhisperToMe, Who, Whouk, WickerWiki, Wik, Wikiacc, Wikibofh, Wikijanko, WikipedianMarlith, Wikipedister, Wikithom, Wikiwonka88, Wilfred Pau, Willdow, William Avery, Witnessdon, Wiz9999, Wmahan, Wombat0121, Woohookitty, Xaronyr, Xcw, Xiaoyu of Yuxi, YA MAM STINKS, Yeanold Viskersenn, YellowMonkey, Yoenit, Yorkshirian, Yosh2008, Zaian, Zanimum, Zaorish, Zarcadia, Zeastop, Zigger, Zoe, Zoicon5, Zouavman Le Zouave, Zulucreative, Zzuuzz, 2663 anonymous edits

List of Acts of the Parliament of the United Kingdom, 1880–1899 *Source*: http://en.wikipedia.org/w/index.php?title=List_of_Acts_of_the_Parliament_of_the_United_Kingdom%2C_1880%E2%80%931899 *Contributors*: Aatomic1, Andrew Gwilliam, Ceyockey, D6, James500, Kurando, Lozleader, Osioni, Patstuart, Sean1989-4, Shimgray, Silverhelm, Starbois, Swiftey127, The Duke of Waltham, Tim!, 7 anonymous edits

Birmingham Corporation Water Department *Source*: http://en.wikipedia.org/w/index.php?title=Birmingham_Corporation_Water_Department *Contributors*: Chris the speller, Martarius, Oosoom, Trident13, 13 anonymous edits

Elan Valley Reservoirs *Source*: http://en.wikipedia.org/w/index.php?title=Elan_Valley_Reservoirs *Contributors*: Alan Liefting, Ascorbic, B, Badgernet, Bahmer, Bardsandwarriors, Blue73, Born Acorn, Bryan Derksen, Cutler, Enaidmawr, Fingerpuppet, Gareth Wyn, Goingin, Grafen, Gwernol, Hydrogen Iodide, JBellis, Jdsplodge, Jeremy Bolwell, Kingboyk, Lee Elms, LinguisticDemographer, Loganberry, Mattisse, Maxburgoyne, Modal Jig, NickBush24, Owain, Paulcardy, Peterb3, Philip Baird Shearer, Pigsonthewing, Pyrotec, Renata, SP-KP, ST47, Safisk, Salvo46, Stemonitis, Velella, 27 anonymous edits

Gravitation *Source*: http://en.wikipedia.org/w/index.php?title=Gravitation *Contributors*: &Delta, 100110100, 24fan24, 2over0, 72Dino, 99DBSIMLR, 9holdss, A Softer Answer, A. di M., A5y, ADNghiem501, AMCKen, APT, Aaron Schulz, Acannas, Accurizer, Adamwang, Addps4cat, Adolphus79, Aechols, Aerographer1981, Aetheling, Ahmediq152, AikBkj, Airplaneman, Al.Glitch, Alan Liefting, Alansohn, Albin&dani, Ale jrb, Alexei Kojenov, AlexiusHoratius, Alexnye, Algont, Allen Jesus, Ally1604, Alphachimp, AnakngAraw, Anaxial, Ance.cdas, Ancheta Wis, Anclation, Andonic, Andrius.v, Androstachys, Andwor9, Andy M. Wang, Andyjsmith, Andykass, AngelOfSadness, Annax3, Anonymous Dissident, Anonymous editor, Another Matt, Antandrus, Anthony Bradbury, Apostolos Margaritis, Applejacks47, Aranherunar, ArkianNWM, Artichoker, Artist7337, Ashanda, Ashmoo, Ask123, AtD, Atif.t2, Atlant, Atwarwiththem, AuburnPilot, Aude, Avono, Azcolvin429, Baconlover13, Ballsucker22, Basawala, BatteryIncluded, Bbq332, Bcasterline, Beetfarm Louie, Beethoven314, Belovedfreak, BenRG, Bender235, Benhocking, Bennybp, Betterusername, Bhadani, Bhahn0125, Bhny, BiT, Bkell, Black Kite, Blackdragon 1002, Blckavnger, Blobbucket, Bloomingiris, Bo Jacoby, Boaex, Bobblehead, Bobblewik, Bobo192, Bobocheese, Bongwarrior, Bookandcoffee, Brandmeister (old), Brandon, BreakfastTom, Brianga, Brownsteve, Bryan Derksen, Bryanmcdonald, Bubba hotep, BuickCenturyDriver, Buttle, C'est moi, C0nanPayne, Caknuck, CalebNoble, Caltas, Caltrop, CambridgeBayWeather, Camw, Can't sleep, clown will eat me, CaptainIraq, CarbonX, Careless hx, Carlosp420, Ccmccm, Cedars, Celebration1981, Cessator, Ch0rx, Chaosmaker39, CharonM72, Charvest, Chase me ladies, I'm the Cavalry, Chase4813, Checkmark56, Chickyfuzz14, Childhoodsend, ChillDeity, Chinneeb, Chives4life, Chris G, Chris earl 89, Chrispasquale, Chs^3, Chun-hian, Cjthellama, ClosedEyesSeeing, Cmichael, Cnormansen, Cockneyite, CodeWeasel, Codingmasters, CombatCraig, Complexica, Coredesat, Cosmictinker, Craigboy, Crazytonyi, Crd721, Cre84u, Cremepuff222, Crunchedfor6, Crustillicus, Cryocide, Cush, Cybercobra, D, D. Recorder, DMacks, DSRH, DVD R W, DVdm, DanMS, DanielCD, Danny, Danski14, Danwills, DarkLink, Darkfrog24, DarlingFriend, Dave-ros, Davecas0, David J Wilson, David Shear, Dbachmann, Delaszk, Delldot, Demiurge, Deor, Deproduction, DerHexer, Dffgd, Dgaubin, Dillard421, Dina, Dirkbb, Discospinster, Dje 8, Djeneba, DocRocks1, Dolphin51, Dominics Fire, Donarreiskoffer, Doulos Christos, Draicone, Drappel, Drdisque, Dreadstar, Drpickem, Dsantesteban, Dylan Lake, Dysepsion, E.vondarkmoor, ERcheck, EamonnPKeane, Eaolson, Ec5618, EdJohnston, Edwardwittenstein, Eedlee, El C, Elembis, Eleuther, Elipongo, Elliott Fontain, Emc2, Ems57fcva, Endomion, Enormousdude, Enviroboy, Epbr123, Epolk, EugeneKantarovich, Eurosong, Everard Proudfoot, Everyking, FJPB, Fairandbalanced, Falcorian, Faradayplank, Fedosin, Feezo, Fel64, Filll, Firth m, Fish and karate, FisherQueen, Fizzackerly, Flamingpanda, Fmyhr, Foobird107, Fplay, Frag1983, FrancoGG, Freak104, Freestyle-69, Frencheigh, Fresheneesz, Fropuff, Fsunka, Funandtrvl, Furkhaocean, GHe, GTBacchus, GVIlleneuve27, Gafferuk, Gakusha, Galo1969X, Gamer007, Garrett gagne, GarrisonGreen, Gatoradeparade, GavinTing, Gdo01, Geoffrey Wickham, Geoffrey.landis, George100, Gershwinrb, Gfoley4, Gholson, GhostPirate, Giftlite, Gimmetrow, Ginkgo100, Ginsengbomb, Glacialvortex, Gladiator2155, Glass Sword, GoldRenet, GoldenGolem, Goldom, GraemeMcRae, Gravityc, Gravityforce, Grayshi, Gregbard, GregorB, Groosh, Grstain, Gsd65, Gseletko, Gtg204y, Guliolopez, Gurch, Gurchzilla, Gwandoya, Gwernol, Haeleth, Hairy Dude, HalfShadow, Hall Monitor, Hallenrn, Hamiltondaniel, Handcannonbeast, Happyhacker101, Hardyplants, Harvardstudent, Hdeasy, Hdt83, Headbomb, Heerojyuy, Heeroyuy135, Hendrixjoseph, Heroesrule17, Heron, Hgrobe, Hi There, Hide&Reason, High2lowo, Hillman, Hisagi, HistoryMaster 1, Hmains, Hmrox, Hollop09, Huntster, Husond, Hut 8.5, Hydrogen Iodide, II MusLiM HyBRiD II, IPSOS, Iain99, IceBlade710, Icestorm815, Icreaser, Igoldste, Inductiveload, Infonation101, Inkling, Inquisitus, InternetMeme, Interwiki gl, Into The Fray, Invmog, Ioannes93, Ishdarian, Isnow, Itel94, Ixfd64, J Di, J.S.B.Anderson, J.delanoy, JDspeeder1, JForget, JHunterJ, JMK, JYolkowski, Jackey0105, Jaganath, Jagged 85, Jakohn, James Mead, Jamesluster, JanSuchy, Jauhienij, JavaJesus, Jaykay0424, Jdiyef, Jebus989, Jeff Relf, Jeff5102, Jennavecia, Jess Mars, Jesvane, Jiddisch, JimJast, Jimip, Jimp, JinJian, Jirachipokemon, Jj1236, Jklin, Jmlk17, Jni, John, John fromer, John254, Jon Stockton, Jonnymagic, Jonpalmer, Joshisamazing, Jossi, Jt16733, Juansidious, JustWong, Jxzj, Kanat Abildinov, Kane1047, Kanwaraj, Karol Langner, Kaseman519, Katrina Graziano, Kazrak, Kazvorpal, Kcin213, Keegan, Keilana, Keith Edkins, Kendelarosa5357, Kenneth M Burke, Kevin, Kicka, Kilo-Lima, King of Hearts, Kiteinthewind, Kitty53, Kmarinas86, KnowledgeOfSelf, Kueller1, Kungfuadam, Kurtan, Kuru, Kwamikagami, L Kensington, LOL, Lahiru k, Lakers, Lappado, Larklight, Larry V, Ld100, Le Docteur, LealandA, Lehkost, LeoNomis, Lesnail, LestatdeLioncourt, Leuko, Lhynard, Liberal Classic, Lightmouse, Ligulem, Linas, Lindberg G Williams Jr, Lithpiperpilot, LittleOldMe old, Logicus, Lolerballer, Loom91, Lorax835, Loren36, LorenzoB, Lost on belmont, Lsskys, LuYiSi, Luk, Lumos3, Luna Santin, Luskj, M C Y 1008, M a s, M keshe, M2Ys4U, MER-C, MJCdetroit, MOBle, MONGO, MPerel, Mac, Mac Davis, MacTire02, Maddox1, Magmafox, Manticore55, MarkSutton, Markp93, MarsRover, Matt Bernius, Matt Deres, MattCarterSurrealist, Maxis ftw, Maya Levy, Mazza uk04, Mccann tom, MeEricYay, MeStevo, Meaningful Username, Meekywiki, Megalobingosaurus, Mejor Los Indios, Mentifisto, Mgiganteus1, Mgnbar, Mhamhamha, Mi6agent00g off, Michael Hardy, MichaelMaggs, Mickeyd24, Micmic28, Mike Rosoft, Mikez, Millahnna, Minna Sora no Shita, Missphysics, Misterbeal, MitchLay, Mitchell is hollywood, Mnkyman, Mohrflies, Mollwollfumble, Momolee, Mononoke, Montrealais, MoogleEXE, Moony1000, Mpatel, Mr Rookles, Mr Stephen, Mr mimises, Mr. Lefty, Mr. dick 008, Mr.Pinklesworth, MrOllie, Mschel, Mtesm, MtyQuinn, Mufka, Muhends, Murgh, Musicpvm, Myagooshki666, Myanw, N shaji, NHRHS2010, NaiPiak, Nakon, Narutolovehinata5, NateDres23, Nathanael Bar-Aur L., NatureA16, NawlinWiki, Nebulosus, Negativecharge, Neil916, Nerval, Nevadacall, Newone, Nicesai, Nichiran, Nick, Nick Number, Nicolas1981, Nihiltres, Ninjainventor, Noctibus, Noel rebeira, Nonsuch, Northgrove, Notinasnaid, Novangelis, Nrbelex, Nuttycoconut, Nyrox395, Nyw195, OMGsplosion, Ocaasi, Oda Mari, Oddball.bfi, Ohwilleke, Ola Hansen, Old Moonraker, Omicronpersei8, Oms22, Opabinia regalis, Optimist on the run, Orion 8, Oroso, Otolemur crassicaudatus, OwenX, Oxymoron83, PSUMark2006, Pablothegreat85, Paleorthid, Panicum, Parvons, Patrick, Patstuart, Pekaje, Persistent76, Phe, Philbarker, Philip Hazelden, Philip Trueman, Phoebe, PhySusie, Piano non troppo, Picaroon, Pieman123456789, Pigman, Pill, Pogogunner, Pollinator, Porterjoh, Ppntori, Presentabsent, PrestonH, Prolog, Pwnasaurusrex38, Pyrospirit, Quarl, Qurq, R'n'B, RB972, RG2, RJaguar3, Radiant!, RadicalOne, Raelx, Raichu Trainer, RainbowOfLight, Raoul NK, Rayc, Raymond arritt, Razorflame, Red Director, Red1, Redmotherfive, Redrocket, Reedy, Rememberway, Rettetast, RexNL, Rhlowe, Riana, Richard Arthur Norton (1958-), Richard L. Peterson, Ridersbydelta, Risker, Rjwilmsi, Rklawton, Rmhermen, Robbie098, RobertFritzius, RobertG, RobertMfromLI, RockSolidCosmo, RogerEllman, Roman à clef, Ronbo76, Roy Brumback, RoyBoy, Rracecarr, Rror, Rsm99833, Ruakh, Rubicon, Ruslik0, Ryan.vilbig, SCZenz, SDC, SFGiants, SJP, SQL, Sadi Carnot, SallyBoseman, Sam42, Samaster1991, Samtheboy, Sander123, Sango123, Saros136, Sarthella, Sbharris, Sbyrnes321, Scarian, Sceptre, Scientizzle, Scm83x, Scohoust, Scot.parker, Scott3, Scott5834, Sct5333, Seba5618, Semperf, Senator Palpatine, Seph Vellius, Seraphim, Sesshomaru, Sevenaces, Shaidar cuebiyar, Shakespeare87, ShakingSpirit, Sheeana, ShelfSkewed, Sheliak, Shenhemu, ShexRox, Shimgray, Shimwell, Shinkolobwe, Shipmaster, Shocky95, SigPig, Silly rabbit, Sin-man, Sinfear, Skateboards fly, Skins88, Skoobieschnax, SkyWalker, Slakr, Smack, Smalljim, Smartman10, Smilesfozwood, Smurrayinchester, Snarfies, Snowolf, Solo Zone, Sophia, Special-T, Speria, Spiel496, Srleffler, Srossman07, Steel, Stephane Yelle, Stephen, StephenBuxton, Steve Pucci, SteveMcCluskey, Steven Weston, Stevertigo, Stoked, Stoned Proffesor, SubwayEater, Supercoop, Supuhstar, Sydweighz, Sylent, Syncategoremata, Sławomir Biały, THEN WHO WAS PHONE?, TJ Elliot Scott, TTE, Tabletop, TakuyaMurata, Tallcreek, Talon Artaine, Tangotango, Tanthalas39, Tardis, Taucetiman, Tbhotch, Techman224, Tecsup, TedE, TedlyW, That Guy, From That Show!, The Blade of the Northern Lights, The One Cause, The Rambling Man, The Random Editor, The Thing That Should Not Be, The Ubik, The monkeyhate, The way, the truth, and the light, TheAlphaWolf, TheRanger, Thefro552, Thegathering, Thejman123456789, Themaxeditor, Thin joe, Thingg, ThomasO1989, ThomasThePolishMan, Thong123456789, Tide rolls, Tiggerjay, Timotheus Canens, TimothyPilgrim, TimothyRias, Tiptoety, Titoxd, Titus III, Tomasz Prochownik, Tonmoy Chowdhury, Tosun, Touch Of Light, Tpbradbury, Travelbird, TravisTX, TreasuryTag, Treisijs, Trentc, Trevor MacInnis, Triwbe, Trusilver, Truthnlove, Ttimespan, TutterMouse, Typesships, Typhoonchaser, U.S.A.U.S.A.U.S.A.,

UNHchabo, Ulterior19802005, UnfriendlyFire, Unicornlad, Utcursch, VCleemput, VandalCruncher, Vanished user, Veddan, Venera 7, Veraptor, Vertigo900, Vorpal blade, Voyagerfan5761, Vrsixfire, Vsmith, WFPM, WHATaintNOcountryIeverHEARDofDOtheySPEAKenglishINwhat, WLU, Wandering Ghost, WarthogDemon, Washboard6, Wavelength, Waxigloo, Wdfarmer, Weregerbil, Where, Whitepaw, Wiki alf, Will Beback, William M. Connolley, Willking1979, Willow123, Willtron, Wimt, Wintran, Witchzilla, Wizardman, Wnt, Wolfkeeper, Work permit, Wtsbeynon, Wvogeler, Xarr, Xerbaycom, Xiahou, Xn4, Yamaguchi, Yay unto the Chicken, Yecril, Yeti Hunter, Yevgeny Kats, Yonatan, Ytcracker, Yuckfoo, Z.E.R.O., Zachary, Zaheen, Zayzya, Zharradan.angelfire, Zureks, , 1796 anonymous edits

Ground-penetrating radar *Source*: http://en.wikipedia.org/w/index.php?title=Ground-penetrating_radar *Contributors*: AGianno, Aatomic1, Adambro, Birczanin, CDN99, Canthusus, Chris the speller, DavidCary, Davidjohndaniels, Dennis Bratland, Eiriksv, Emulas, Eve Hall, Exp4000, Fcueto, Fratrep, Ftsisthebest, Fulldecent, GSSI World Leader, Geoinstberlin, Georadar, Geowizd, Gnowa7, GregorB, Grika, Hesperian, Hilaryburt14, Immo trinks, InfrasenseInc, JamesBWatson, JeffreyN, Jiuguang Wang, Jn52, Joemacgregor, Jonas Matuzas, Jstuby, Locris, MC10, MGlosenger, Maayanh, Macs4life, Mangatron, Mgummess, Mild Bill Hiccup, Mimihitam, Montell 74, Mwtoews, Nonarina, Oatmeal batman, Olivier, PITACL1368, PenguiN42, Pierre cb, Piotrus, Quantpole, Rainspeaker, STHayden, Shiny green, Skysmith, Tapatio, The Thing That Should Not Be, TheGerm, Tommike125, Trevor Marron, 102 anonymous edits

Image Sources, Licenses and Contributors

Image:Taiwan JungHua Dam.JPG *Source*: http://en.wikipedia.org/w/index.php?title=File:Taiwan_JungHua_Dam.JPG *License*: unknown *Contributors*: User:Vegafish

Image:Lakevyrnwysummer.jpg *Source*: http://en.wikipedia.org/w/index.php?title=File:Lakevyrnwysummer.jpg *License*: unknown *Contributors*: User:Sean the Spook. Original uploader was Sean the Spook at en.wikipedia

Image:Stocks Reservoir.jpg *Source*: http://en.wikipedia.org/w/index.php?title=File:Stocks_Reservoir.jpg *License*: unknown *Contributors*: User:Karlos87

Image:GibsonR.jpg *Source*: http://en.wikipedia.org/w/index.php?title=File:GibsonR.jpg *License*: unknown *Contributors*: User:Citypeek

Image:Hydroelectric dam.svg *Source*: http://en.wikipedia.org/w/index.php?title=File:Hydroelectric_dam.svg *License*: unknown *Contributors*: User:Tomia

Image:KupferbachStauseeAachen.jpg *Source*: http://en.wikipedia.org/w/index.php?title=File:KupferbachStauseeAachen.jpg *License*: unknown *Contributors*: User:Túrelio

Image:Llyn Brianne spillway.jpg *Source*: http://en.wikipedia.org/w/index.php?title=File:Llyn_Brianne_spillway.jpg *License*: unknown *Contributors*: User:Velela

Image:volta lake.jpg *Source*: http://en.wikipedia.org/w/index.php?title=File:Volta_lake.jpg *License*: unknown *Contributors*: NASA

Image:Lake Kariba.jpg *Source*: http://en.wikipedia.org/w/index.php?title=File:Lake_Kariba.jpg *License*: unknown *Contributors*: Martin H., Mehmet Karatay, Paobac, TheDJ, TommyBee, 1 anonymous edits

Image:Drinking water.jpg *Source*: http://en.wikipedia.org/w/index.php?title=File:Drinking_water.jpg *License*: unknown *Contributors*: Photo taken by using a Nikon Coolpix 950.

File:Bundesarchiv Bild 101I-784-0233-03, Nordafrika, Wasserversorgungsstelle (-).jpg *Source*: http://en.wikipedia.org/w/index.php?title=File:Bundesarchiv_Bild_101I-784-0233-03,_Nordafrika,_Wasserversorgungsstelle_(-).jpg *License*: unknown *Contributors*: Tannenberg

Image:Mwamongu water source.jpg *Source*: http://en.wikipedia.org/w/index.php?title=File:Mwamongu_water_source.jpg *License*: unknown *Contributors*: Bob Metcalf

Image:E8661-Pattaya-water-vending-machines.jpg *Source*: http://en.wikipedia.org/w/index.php?title=File:E8661-Pattaya-water-vending-machines.jpg *License*: unknown *Contributors*: User:Vmenkov

Image:Water quality.jpg *Source*: http://en.wikipedia.org/w/index.php?title=File:Water_quality.jpg *License*: unknown *Contributors*: User:Ionut Cojocaru

Image:Indonesia-sodis-gross.jpg *Source*: http://en.wikipedia.org/w/index.php?title=File:Indonesia-sodis-gross.jpg *License*: unknown *Contributors*: User:SODIS Eawag

File:Drinking Water Security Poster EPA.jpg *Source*: http://en.wikipedia.org/w/index.php?title=File:Drinking_Water_Security_Poster_EPA.jpg *License*: unknown *Contributors*: Smallman12q

File:BirminghamMontage.jpg *Source*: http://en.wikipedia.org/w/index.php?title=File:BirminghamMontage.jpg *License*: unknown *Contributors*: User:JimmyGuano

File:Birm 1977 arms.png *Source*: http://en.wikipedia.org/w/index.php?title=File:Birm_1977_arms.png *License*: unknown *Contributors*: User:OrphanBot

File:Birmingham UK locator map.svg *Source*: http://en.wikipedia.org/w/index.php?title=File:Birmingham_UK_locator_map.svg *License*: unknown *Contributors*: User:Nilfanion

File:Flag of the United Kingdom.svg *Source*: http://en.wikipedia.org/w/index.php?title=File:Flag_of_the_United_Kingdom.svg *License*: unknown *Contributors*: Anomie, Mifter

File:Flag of England.svg *Source*: http://en.wikipedia.org/w/index.php?title=File:Flag_of_England.svg *License*: unknown *Contributors*: Anomie

Image:Speakerlink.svg *Source*: http://en.wikipedia.org/w/index.php?title=File:Speakerlink.svg *License*: unknown *Contributors*: Woodstone. Original uploader was Woodstone at en.wikipedia

File:Birmingham - Westley Map 1731.png *Source*: http://en.wikipedia.org/w/index.php?title=File:Birmingham_-_Westley_Map_1731.png *License*: unknown *Contributors*: Original uploader was JimmyGuano at en.wikipedia

File:Birmingham in 1886.jpg *Source*: http://en.wikipedia.org/w/index.php?title=File:Birmingham_in_1886.jpg *License*: unknown *Contributors*: Original uploader was G-Man at en.wikipedia

File:Victoria Square, Birmingham at dusk.jpg *Source*: http://en.wikipedia.org/w/index.php?title=File:Victoria_Square,_Birmingham_at_dusk.jpg *License*: unknown *Contributors*: Cristian Bortes

File:Birmingham panorama from the Lickey Hills.jpg *Source*: http://en.wikipedia.org/w/index.php?title=File:Birmingham_panorama_from_the_Lickey_Hills.jpg *License*: unknown *Contributors*: User:JimmyGuano

File:Birmingham canalside apartments at dusk.jpg *Source*: http://en.wikipedia.org/w/index.php?title=File:Birmingham_canalside_apartments_at_dusk.jpg *License*: unknown *Contributors*: ...some guy

File:ColmoreRowBirmingham.jpg *Source*: http://en.wikipedia.org/w/index.php?title=File:ColmoreRowBirmingham.jpg *License*: unknown *Contributors*: User:GavinWarrins

File:Jaguar XJ X351.jpg *Source*: http://en.wikipedia.org/w/index.php?title=File:Jaguar_XJ_X351.jpg *License*: unknown *Contributors*: TTTNIS

File:Black Sabbath 1999-12-16 Stuttgart.jpg *Source*: http://en.wikipedia.org/w/index.php?title=File:Black_Sabbath_1999-12-16_Stuttgart.jpg *License*: unknown *Contributors*: User:Alex Ex

File:ExCathedra-BirminghamTownHall-byJamesAshby-20080301.jpg *Source*: http://en.wikipedia.org/w/index.php?title=File:ExCathedra-BirminghamTownHall-byJamesAshby-20080301.jpg *License*: unknown *Contributors*: James Ashby.

File:Birmingham Hippodrome.jpg *Source*: http://en.wikipedia.org/w/index.php?title=File:Birmingham_Hippodrome.jpg *License*: unknown *Contributors*: Original uploader was Oosoom at en.wikipedia

File:David Cox - Rhyl Sands (Tate version).jpg *Source*: http://en.wikipedia.org/w/index.php?title=File:David_Cox_-_Rhyl_Sands_(Tate_version).jpg *License*: unknown *Contributors*: David Cox (1783–1859)

File:BirminghamBarberInstitute.jpg *Source*: http://en.wikipedia.org/w/index.php?title=File:BirminghamBarberInstitute.jpg *License*: unknown *Contributors*: User:GavinWarrins

File:B'ham academy.jpg *Source*: http://en.wikipedia.org/w/index.php?title=File:B'ham_academy.jpg *License*: unknown *Contributors*: Loudrocksurfer at en.wikipedia

File:Bell Edison Telephone Building.jpg *Source*: http://en.wikipedia.org/w/index.php?title=File:Bell_Edison_Telephone_Building.jpg *License*: unknown *Contributors*: Original uploader was Oosoom at en.wikipedia

File:Blob Birmingham.jpg *Source*: http://en.wikipedia.org/w/index.php?title=File:Blob_Birmingham.jpg *License*: unknown *Contributors*: User:Bs0u10e01

File:Birmingham airport arrivals arp.jpg *Source*: http://en.wikipedia.org/w/index.php?title=File:Birmingham_airport_arrivals_arp.jpg *License*: unknown *Contributors*: Arpingstone, 2 anonymous edits

File:390029 'City of Stoke-on-Trent' at Birmingham New Street.JPG *Source*: http://en.wikipedia.org/w/index.php?title=File:390029_'City_of_Stoke-on-Trent'_at_Birmingham_New_Street.JPG *License*: unknown *Contributors*: User:Our Phellap

File:BirminghamUniversityChancellorsCourt.jpg *Source*: http://en.wikipedia.org/w/index.php?title=File:BirminghamUniversityChancellorsCourt.jpg *License*: unknown *Contributors*: User:GavinWarrins

File:BirminghamCentralLibrary.jpg *Source*: http://en.wikipedia.org/w/index.php?title=File:BirminghamCentralLibrary.jpg *License*: unknown *Contributors*: Original uploader was SilentVicinity at en.wikipedia

File:Charles Gore - Statue - St. Philip's - Birmingham - 2005-10-14.jpg *Source*: http://en.wikipedia.org/w/index.php?title=File:Charles_Gore_-_Statue_-_St._Philip's_-_Birmingham_-_2005-10-14.jpg *License*: unknown *Contributors*: Kilom691, Man vyi, Oosoom, Tagishsimon

File:AstonvillavsBirmingham06.JPG *Source*: http://en.wikipedia.org/w/index.php?title=File:AstonvillavsBirmingham06.JPG *License*: unknown *Contributors*: Original uploader was Mozman at en.wikipedia

File:Edgbaston - view of new stand from the north.jpg *Source*: http://en.wikipedia.org/w/index.php?title=File:Edgbaston_-_view_of_new_stand_from_the_north.jpg *License*: unknown *Contributors*: User:JimmyGuano

File:NationalIndoorArena.jpg *Source*: http://en.wikipedia.org/w/index.php?title=File:NationalIndoorArena.jpg *License*: unknown *Contributors*: Cls14 (talk). Original uploader was Cls14 at en.wikipedia

File:Birmingham Wholesale Markets.jpg *Source*: http://en.wikipedia.org/w/index.php?title=File:Birmingham_Wholesale_Markets.jpg *License*: unknown *Contributors*: star-one

File:ElectricCinema.jpg *Source*: http://en.wikipedia.org/w/index.php?title=File:ElectricCinema.jpg *License*: unknown *Contributors*: SilentVicinity. Original uploader was SilentVicinity at en.wikipedia

File:Mailbox at Night.jpg *Source*: http://en.wikipedia.org/w/index.php?title=File:Mailbox_at_Night.jpg *License*: unknown *Contributors*: User:Bs0u10e01

File:Matthew Boulton - Carl Frederik von Breda.jpg *Source*: http://en.wikipedia.org/w/index.php?title=File:Matthew_Boulton_-_Carl_Frederik_von_Breda.jpg *License*: unknown *Contributors*: Original uploader was JimmyGuano at en.wikipedia

File:Flag of the United States.svg *Source*: http://en.wikipedia.org/w/index.php?title=File:Flag_of_the_United_States.svg *License*: unknown *Contributors*: Anomie

File:Flag of Germany.svg *Source*: http://en.wikipedia.org/w/index.php?title=File:Flag_of_Germany.svg *License*: unknown *Contributors*: Anomie

File:Flag of South Africa.svg *Source*: http://en.wikipedia.org/w/index.php?title=File:Flag_of_South_Africa.svg *License*: unknown *Contributors*: Adriaan, Anime Addict AA, AnonMoos, BRUTE, Daemonic Kangaroo, Dnik, Duduziq, Dzordzm, Fry1989, Homo lupus, Jappalang, Juliancolton, Kam Solusar, Klemen Kocjancic, Klymene, Lexxyy, Mahahahaneapneap, Manuelt15, Moviedefender, NeverDoING, Ninane, Poznaniak, Przemub, SKopp, ThePCKid, ThomasPusch, Tvdm, Ultratomio, Vzb83, Zscout370, 33 anonymous edits

File:Flag of France.svg *Source*: http://en.wikipedia.org/w/index.php?title=File:Flag_of_France.svg *License*: unknown *Contributors*: Anomie

File:Flag of Italy.svg *Source*: http://en.wikipedia.org/w/index.php?title=File:Flag_of_Italy.svg *License*: unknown *Contributors*: Anomie

File:Edgbaston Waterworks.jpg *Source*: http://en.wikipedia.org/w/index.php?title=File:Edgbaston_Waterworks.jpg *License*: unknown *Contributors*: User Oosoom on en.wikipedia

File:Foel Tower Elan Valley - geograph.org.uk - 145135.jpg *Source*: http://en.wikipedia.org/w/index.php?title=File:Foel_Tower_Elan_Valley_-_geograph.org.uk_-_145135.jpg *License*: unknown *Contributors*: Docu

File:Pery Barr Reservoir CoB sign.JPG *Source*: http://en.wikipedia.org/w/index.php?title=File:Pery_Barr_Reservoir_CoB_sign.JPG *License*: unknown *Contributors*: -

Image:Caban Coch, Elan Valley.jpg *Source*: http://en.wikipedia.org/w/index.php?title=File:Caban_Coch,_Elan_Valley.jpg *License*: unknown *Contributors*: Docu, Loganberry, RHaworth, Ronaldino, Victuallers

Image:Caban Goch-1-.jpg *Source*: http://en.wikipedia.org/w/index.php?title=File:Caban_Goch-1-.jpg *License*: unknown *Contributors*: User:Maxburgoyne

Image:Garreg ddu panorama.jpg *Source*: http://en.wikipedia.org/w/index.php?title=File:Garreg_ddu_panorama.jpg *License*: unknown *Contributors*: T Johnson (Jdsplodge)

Image:Elanvalley2004.JPG *Source*: http://en.wikipedia.org/w/index.php?title=File:Elanvalley2004.JPG *License*: unknown *Contributors*: Blue73, 1 anonymous edits

Image:Solar sys.jpg *Source*: http://en.wikipedia.org/w/index.php?title=File:Solar_sys.jpg *License*: unknown *Contributors*: Harman Smith and Laura Generosa (nee Berwin), graphic artists and contractors to NASA's Jet Propulsion Laboratory.

Image:Falling ball.jpg *Source*: http://en.wikipedia.org/w/index.php?title=File:Falling_ball.jpg *License*: unknown *Contributors*: User:MichaelMaggs

Image:GalacticRotation2.svg *Source*: http://en.wikipedia.org/w/index.php?title=File:GalacticRotation2.svg *License*: unknown *Contributors*: User:PhilHibbs

Image:LINE21.jpg *Source*: http://en.wikipedia.org/w/index.php?title=File:LINE21.jpg *License*: unknown *Contributors*: Original uploader was Tapatio at en.wikipedia

File:Ground Penetrating Radar in use.jpg *Source*: http://en.wikipedia.org/w/index.php?title=File:Ground_Penetrating_Radar_in_use.jpg *License*: unknown *Contributors*: Bdk, Blonder1984, FlickreviewR, Macs4life

Image:Radarsurvey.jpg *Source*: http://en.wikipedia.org/w/index.php?title=File:Radarsurvey.jpg *License*: unknown *Contributors*: Lradrama, Tapatio, Timichal, 1 anonymous edits

MIX
Papier aus verantwortungsvollen Quellen
Paper from responsible sources
FSC® C105338

Printed by Books on Demand GmbH, Norderstedt / Germany